GIS and Spatial Analysis

Edited by Jorge Rocha, Eduardo Gomes, Inês Boavida-Portugal, Cláudia M. Viana, Linh Truong-Hong and Anh Thu Phan

Published in London, United Kingdom

GIS and Spatial Analysis
http://dx.doi.org/10.5772/intechopen.100705
Edited by Jorge Rocha, Eduardo Gomes, Inês Boavida-Portugal, Cláudia M. Viana, Linh Truong-Hong and Anh Thu Phan

Contributors
Efthymios-Spyridon Georgiou, Alfred Homère Ngandam Mfondoum, Igor Casimir Njombissie Petcheu, Frederic Chamberlain Lounang Tchatchouang, Luc Moutila Beni, Mesmin Tchindjang, Jean Valery Mefire Mfondoum, Sevgi Eda Tuzcu, Esra Satıcı, Jonas Didero Takodjou Wambo, Wenzi Tanni Quinter, Brooks C. C Pearson, Feleke Asrat, Matthew H. Connolly, Gizachew Tiruneh, Santiago Royo, Maria Ballesta-Garcia, Gerard DeMas-Giménez, Yongqin Zhang, Kory Iman, Fang Yong, Cao Bincai, Gong Hui, Gao Li, Zhang Li, Hu Haiyan, Abetov Auez Egemberdievich, Yessirkepova Sharbanu Bakhytovna, Julia Barbosa Curto Ma, Jorge Rocha, Cláudia M. Viana, Inês Boavida-Portugal, Eduardo Gomes

First published in London, United Kingdom, 2023 by IntechOpen
IntechOpen is the global imprint of INTECHOPEN LIMITED, registered in England and Wales, registration number: 11086078, 5 Princes Gate Court, London, SW7 2QJ, United Kingdom

British Library Cataloguing-in-Publication Data
A catalogue record for this book is available from the British Library

Additional hard and PDF copies can be obtained from orders@intechopen.com

GIS and Spatial Analysis
Edited by Jorge Rocha, Eduardo Gomes, Inês Boavida-Portugal, Cláudia M. Viana, Linh Truong-Hong and Anh Thu Phan
p. cm.
Print ISBN 978-1-80356-596-5
Online ISBN 978-1-80356-597-2
eBook (PDF) ISBN 978-1-80356-598-9

For EU product safety concerns:
IN TECH d.o.o., Prolaz Marije Krucifikse Kozulić 3, 51000 Rijeka, Croatia,
info@intechopen.com or visit our website at intechopen.com.

Meet the editors

Dr. Jorge Rocha obtained an MSc in Geographic Information Systems and an MSc in Spatial Planning in 2003 and 2013, respectively, as well as a Ph.D. in Geographic Information Science in 2012. He is currently an assistant professor at the Institute of Geography and Spatial Planning and a member of the Modelling, Urban and Regional Planning, and Environmental Hazard and Risk Assessment and Management research groups of the Centre of Geographical Studies, all from the University of Lisbon (ULisboa). His field of expertise is geosimulation and geocomputation involving artificial neural networks, graphs theory, cellular automata, and multi-agent systems. Dr. Rocha's work focuses on urban morphology, remote sensing, land use/land cover, epidemiology, health geography, geomarketing, tourism, smart cities, and Big Data.

Dr. Linh Truong-Hong is a researcher in the Department of Geoscience and Remote Sensing, Delft University of Technology, Netherlands. After completing a Marie Curie Fellowship at the same university, he obtained a BEng and MEng from the Ho Chi Minh City University of Technology, Vietnam, and a Ph.D. from University College Dublin, Ireland. He was a senior researcher at University College Dublin (2013–2018) after finishing his post-doctoral research at the Dublin Institute of Technology (2011–2013). His research interests are to develop integrated frameworks to deploy light detection and ranging (LiDAR) and photogrammetry in 3D objection reconstruction and change detection for infrastructure inspection and assessment, construction management, and a smart city. His research activity has been recognized with six peer-reviewed international awards.

Dr. Eduardo Gomes obtained a Ph.D. in Geography from the University of Paris 1 Panthéon-Sorbonne, France and University of Lisbon (ULisboa). His latest research work focused on the design of modeling and collaborative simulations employed to analyze land use and land use cover changes. He was an invited lecturer at Beijing Forest University, Institut des Systèmes Complexes de Paris, Future Cities Laboratory and at the Complexity Institute, Singapore; University of Copenhagen, Denmark; Royal Melbourne Institute of Technology, Australia. Dr. Gomes is the recipient of several awards, such as the Portuguese National Geography Award for the best Ph.D. thesis, Best Scientific Article Award in Human Geography, and outstanding contributions in reviewing scientific journals. He is currently an assistant professor at the University of Lisbon.

Dr. Cláudia M. Viana is a geospatial data scientist and a geographer. She obtained a master's degree in Geographic Information Systems in 2014 and a Ph.D. in Geographic Information Science in 2022. She is currently a data science junior researcher at the Centre of Geographical Studies and Institute of Geography and Spatial Planning, University of Lisbon (ULisboa), and a member of the Modelling and Spatial Planning research group. Dr. Viana was a fellow researcher in several scientific projects related to changes in agriculture and the link between transport and land use at the Institute of Social Sciences and the Faculty of Architecture, both at ULisboa. Her research deals with the spatiotemporal changes in agricultural land and its implications on food security and health.

Dr. Inês Boavida-Portugal is an assistant professor at the University of Lisbon's Institute of Geography and Spatial Planning (IGOT-ULisboa). Her recent work explores urban tourism patterns and spatial planning responses to gear cities towards future sustainable pathways. To explore these phenomena, Dr. Boavida-Portugal uses a mixed-method approach including geographical information systems and information and communication technologies. She has been particularly interested in the application of such methodologies to study how behavioral changes at the micro level (governance triangle: civil society, governments, markets) can produce critical changes at the macro level in urban systems, specifically focusing on tourism dynamics as a disruptive phenomenon.

Dr. Anh Thu Phan is a lecturer at Vietnam National University (VNU), Ho Chi Minh City University of Technology. She finished her undergraduate education as a salutatorian and was in the top 10 students of her faculty at VNU. After that, she spent two years working at VNU as an assistant lecturer. She obtained an MSc from Nagaoka University of Technology, Japan, in 2014 where she had a full academic scholarship. She obtained a Ph.D. from the same university in 2017. Since then, she has been a full-time lecturer at the Department of Geomatics Engineering, Faculty of Civil Engineering, Ho Chi Minh City University of Technology. Her research focuses on the application of low-cost laser scanners and closed-range photogrammetry.

Contents

Preface

Spatial analysis and geographic information are interconnected themes that complement each other in the study and understanding of space and territory, allowing, among other things, to analyze and understand spatial patterns and territorial dynamics of phenomena and processes.

Geographic information is the basis for spatial analysis. Its quality is a critical aspect of making informed decisions and obtaining reliable results. It is essential to ensure the precision, accuracy, reliability, completeness, consistency, timeliness, readability, and comprehensibility of the geographic data used. This can be achieved through proper data collection, validation, standardization, and updating techniques.

In this context, the increasing availability of Big Data (structured or not) has allowed obtaining more detailed representations of human behavior and movement patterns, environmental changes, and other complex spatial dynamics. The seven Vs (Volume, Variety, Velocity, Variability, Veracity, Visualization, and Value) enabled the creation of more accurate complex models and the understanding of phenomena that were previously difficult to analyze at different scales.

Through the complex systems approach, it is possible to understand the interactions and feedback between the elements that make up space and territory and how these interactions influence spatial patterns and territorial dynamics. The spatially explicit models, with special emphasis on those supported by artificial intelligence (where machine learning fits), help the understanding and prediction of spatiotemporal dynamics, allowing us to simulate future and/or past scenarios, identify patterns, and test hypotheses. Its areas of application are quite diverse, covering themes such as urban expansion, the spread of diseases, the distribution of natural resources, and transport and mobility, among others, as long as they have a spatial dimension.

Thus, with the simultaneous emergence of new data sources and new methods of analysis, new concepts and tools have emerged, such as the smart city, circularity and urban metabolism, mixed models (qualitative/quantitative), multiverse, agnostics and agents, virtual and augmented reality, digital twins, and the metaverse. Despite being different in substance and area of application, all these approaches aim to support the process of territorial analysis, allowing for understanding the organization and transformations of space and identifying the interdependent relationships between the different elements that compose it. The territorial analysis is fundamental for understanding spatial inequalities and for adequate planning and management of the territory, which is essential for supporting decision-making.

In short, spatial analysis and modeling cover a wide range of themes and topics, as the theme is interdisciplinary and comprises several domains of Geographic Information Science (GISc). As such, this book about spatial analysis in Geographic Information

Systems (GIS) is a powerful tool, as it is a comprehensive resource for students, researchers, and practitioners who are interested in learning about spatial analysis and its applications in GIS.

This book covers a wide range of topics, including the fundamentals of GIS and spatial data, various types of spatial analysis techniques, and their applications in different fields. It also provides examples of how GIS and spatial analysis have been used to solve real-world problems, such as exploring new data sources (e.g., light detection and ranging [LiDar]), analyzing health and environment relationships, mapping land use/land cover changes, modeling transports and infrastructures, or assessing and monitoring the impacts of mining activity.

This volume is a valuable resource for anyone who wants to learn more about spatial analysis in GIS. It provides a clear and concise explanation of complex concepts as well as practical guidance on how to use GIS and spatial analysis tools effectively. Included case studies and examples help readers apply what they have read to develop their own skills in spatial analysis. We hope this book contributes to the field and helps others learn about the power and potential of spatial analysis.

Jorge Rocha, Cláudia M. Viana, Inês Boavida-Portugal and Eduardo Gomes
Institute of Geography and Spatial Planning,
University of Lisbon,
Lisbon, Portugal

Associate Laboratory TERRA,
Lisbon, Portugal

Dr. Linh Truong-Hong
Department of Geoscience and Remote Sensing,
Delft University of Technology,
Delft, Netherlands

Dr. Anh Thu Phan
Vietnam National University (VNU),
Ho Chi Minh City University of Technology,
Ho Chi Minh, Vietman

Section 1

General Insights

Chapter 1

Introductory Chapter: GIS and Spatial Analysis

Cláudia M. Viana, Inês Boavida-Portugal, Eduardo Gomes and Jorge Rocha

1. Introduction

Geographic Information Systems (GIS) and spatial analysis are considered to be a science in their own right, with a solid theoretical and methodological basis. The science behind GIS and spatial analysis has been coined as geoinformatics, which is defined as the application of Geographic Information Science (GISc) to solve problems in earth and environmental sciences. Geoinformatics involves the collection, storage, processing, analysis, visualization, and dissemination of geographic information.

Spatial analysis is a fundamental aspect of geoinformatics and is used to study the distribution and relationship between geographic objects and events. Spatial analysis involves the use of statistical, mathematical, and computational techniques to explore patterns and trends in geographic data. It also allows users to create spatial models and make predictions based on different scenarios.

The science behind spatial analysis involves the application of mathematical, statistical, and computational methods to analyze and interpret spatial patterns and relationships between geographic objects and events. It draws on a variety of disciplines such as geography, mathematics, statistics, computer science, and remote sensing to provide a comprehensive understanding of spatial data.

The theoretical basis of spatial analysis includes concepts such as spatial autocorrelation, spatial heterogeneity, and spatial dependence, which helps to explain the spatial patterns and relationships observed in geographic data. Spatial analysis methods can be broadly categorized into descriptive, exploratory, and inferential techniques, which are used to visualize, explore, and test spatial data.

Some common spatial analysis techniques include spatial interpolation, spatial regression, spatial clustering, spatial smoothing, and spatial econometrics. These methods can be applied to a wide range of spatial data, including point data, areal data, and network data.

Spatial analysis has become increasingly important in many fields such as public health, environmental studies, urban planning, and criminology, among others. It provides a powerful tool to study spatial problems and make informed decisions based on spatial data. Advances in technology have also led to the development of new spatial analysis methods, such as machine learning and deep learning, which are being applied to address complex spatial problems.

IntechOpen

2. Applications

Geoinformatics and spatial analysis have evolved rapidly in recent decades, thanks to emerging technologies such as cloud computing, artificial intelligence, machine learning, and big data. These technologies are making geoinformatics more accessible, powerful, and relevant than ever before. As a result, geoinformatics and spatial analysis are becoming increasingly interdisciplinary, and their applications are expanding into new fields such as healthcare, retail, entertainment.

2.1 Health and environment

GIS and spatial analysis are being increasingly used in health and environmental fields to understand the spatial distribution of diseases, environmental hazards, and their associated risk factors.

In the health field, GIS can be used to map disease incidence and prevalence rates, identify patterns and clusters of diseases [1], and assess the impact of environmental factors on health outcomes [2]. For example, GIS can be used to identify areas with high rates of cancer and investigate whether there are any environmental factors, such as air pollution or water contamination, which may be contributing to the high incidence rates [3–5].

Similarly, in the environmental field, GIS is useful on mapping the distribution of environmental hazards, such as toxic waste sites or air pollution sources, and assess their potential impact on human health. GIS can help identifying areas at risk of natural disasters, such as floods or wildfires, and support emergency response efforts.

Spatial analysis techniques, such as spatial autocorrelation and cluster analysis, help to identify patterns and trends in health and environmental data. For example, spatial autocorrelation can be used to identify areas with similar health outcomes or environmental hazards, while cluster analysis can be used to identify areas with statistically significant clusters of disease or environmental hazards [1].

2.2 Land use land cover changes

Also in land use, land cover change analysis, GIS, and spatial analysis reveal themselves as powerful tools. GIS allows to create maps in order to visualize changes in land use and land cover over time. This allows to identify areas that have undergone significant changes and areas that have remained relatively stable.

Identifying the drivers of land use and land cover changes, such as urbanization, agriculture, or natural disasters, is another field of application. This can help inform land management and policy decisions [6].

Moreover, one can use GIS to create predictive models of future land use and land cover changes, which allows to detect areas that are at risk of change and inform planning and management decisions [7]. We can also assess the impacts of land use and land cover changes on the environment, such as changes in water quality, soil erosion, or biodiversity loss.

Finally, monitoring and tracking changes in land use and land cover over time let to identifying areas that are undergoing rapid changes and inform management decisions.

2.3 Transports and infrastructures

Spatial analysis and GIS are essential tools in the planning, design, and management of transportation and infrastructure systems. Modeling and analyzing transportation networks, such as road networks, transit systems, and bike lanes, are a key application. It helps to optimize routes, identify chokepoints, and plan for future growth.

In addition, they have an important role into infrastructure assets management, such as bridges, tunnels, and pipelines. This can help to identify maintenance needs, track inspections, and plan for replacements.

Another common use is for identifying suitable locations for new infrastructure projects, such as highways, airports, and transit stations, minimizing the environmental impacts and optimizing the use of resources. Furthermore, one can model and analyze the environmental impacts of those infrastructure projects, such as air and water pollution, noise pollution, and habitat destruction. This approach enables to mitigate the impacts and ensure compliance with regulations.

Managing and responding to emergencies, such as natural disasters, traffic accidents, and power outages, are one of the major GIS applications. It reinforces public safety and minimizes the impact of these events [8].

2.4 Mining exploration and monitoring

GIS allows for the efficient management and analysis of large amounts of spatial data, which is critical in the mining industry. GIS technology is well adapted to create detailed maps of mining sites, showing the location of ore deposits, infrastructure, and other important features. This information is further used to plan and manage mining operations, as well as to identify potential areas for further exploration [9].

Spatial analysis techniques, such as geostatistics, help to analyze and model mining data, including geologic and geochemical data, to identify patterns and trends. These analyses can help to optimize the location of mining operations, reduce costs, and increase the accuracy of mineral resource estimates.

In addition, GIS and spatial analysis are fitted for environmental monitoring, such as tracking changes in vegetation, water quality, and air quality around mining sites, ensuring that mining operations are conducted in an environmentally responsible manner and to mitigate any negative impacts.

2.5 Lidar applications

Lidar (Light Detection and Ranging) is a remote-sensing technology that uses lasers to measure distances to the Earth's surface and creates highly accurate 3D models of landscapes and other features [10].

GIS can be used to manage and analyze Lidar data, which can be in the form of point clouds or raster data. Point clouds are collections of 3D points that represent the surface of the Earth, while raster data are a grid of cells that represent the elevation of the Earth's surface. GIS software can process and display both types of data, allowing for analysis and visualization of Lidar-derived information.

Spatial analysis techniques, such as terrain analysis, viewshed analysis, and slope analysis, permit to extract valuable information from Lidar data. For example, terrain analysis makes possible to identify areas of high and low elevation, while viewshed analysis determines the areas that are visible from a certain point. Slope analysis highlights the areas with steep slopes, which can be important for identifying areas prone to landslides or other hazards.

Additionally, GIS can be used to integrate Lidar data with other types of spatial data, such as satellite imagery, demographic data, and land use data [11]. This can provide a more complete picture of the studied areas and allow for more informed decision-making.

3. Conclusions

Spatial data are difficult to interpret by themselves. Turning them into maps and graphs makes them easier to observe and find any patterns. The maps are especially useful because, in addition to the visual component, which facilitates observation, they allow us to overlay different types of information (from aerial photographs, satellite images, and statistical data).

Often, looking at a map is enough to find a distribution pattern or a relationship between variables and their spatial distribution. Combine different groups of data and looking at them from different perspectives (scales) can be a valuable method, and information technology allows to do it quickly.

Lidar data is one of the most recent sources of information and can be used for a wide range of applications, including natural resource management, urban planning, and disaster response.

GIS and spatial analysis deliver valuable tools for understanding the complex relationships between health and the environment and can support evidence-based decision-making in public health and environmental policy.

By providing detailed and accurate information about land use and land cover changes, they can help inform policy and management decisions to ensure sustainable land use practices.

Spatial analysis and GIS are also powerful tools that can help to improve the efficiency, safety, and sustainability of transportation and infrastructure systems and the efficiency, accuracy, and sustainability in the mining industry.

Author details

Cláudia M. Viana[1,2], Inês Boavida-Portugal[1,2], Eduardo Gomes[1,2] and Jorge Rocha[1,2*]

1 Institute of Geography and Spatial Planning, University of Lisbon, Lisbon, Portugal

2 Associate Laboratory TERRA, Lisbon, Portugal

*Address all correspondence to: jorge.rocha@campus.ul.pt

References

[1] Silva M, Betco I, Capinha C, Roquette R, Viana CM, Rocha J. Spatiotemporal dynamics of COVID-19 infections in mainland Portugal [internet]. Sustainability. 2022;**14**(16):10370. DOI: 10.3390/su141610370

[2] Barbosa B, Silva M, Capinha C, Garcia RAC, Rocha J. Spatial correlates of COVID-19 first wave across continental Portugal. Geospatial Health. 2022;**17**(s1). [Internet]. DOI: 10.4081/gh.2022.1073. Available from: https://geospatialhealth.net/index.php/gh/article/view/1073

[3] Roquette R, Painho M, Nunes B. Geographical patterns of the incidence and mortality of colorectal cancer in mainland Portugal municipalities (2007-2011). BMC Cancer. 2019;**19**:512. [Internet]. DOI: 10.1186/s12885-019-5719-9

[4] Cardoso D, Painho M, Roquette R. A geographically weighted regression approach to investigate air pollution effect on lung cancer: A case study in Portugal. Geospatial Health. 2019;**14**(1). [Internet]. DOI: 10.4081/gh.2019.701. Available from: https://www.geospatialhealth.net/index.php/gh/article/view/701

[5] Roquette R, Painho M, Nunes B. Spatial epidemiology of cancer: A review of data sources, methods and risk factors. Geospatial Health. 2017;**12**(1). [Internet]. DOI: 10.4081/gh.2017.504. Available from: https://www.geospatialhealth.net/index.php/gh/article/view/504

[6] Viana CM, Santos M, Freire D, Abrantes P, Rocha J. Evaluation of the factors explaining the use of agricultural land: A machine learning and model-agnostic approach. Ecological Indicators. 2021;**131**:108200 [Internet]. Available from: https://www.sciencedirect.com/science/article/pii/S1470160X21008657

[7] Gomes E, Banos A, Abrantes P, Rocha J, Schläpfer M. Future land use changes in a peri-urban context: Local stakeholder views. Science in the Total Environment. 2020;**718**:137381 [Internet] Elsevier; [cited 2020 Feb 27] Available from: https://www.sciencedirect.com/science/article/pii/S0048969720308913?via%3Dihub

[8] Melo R, Zêzere JL, Oliveira SC, Garcia RAC, Oliveira S, Pereira S, et al. Defining evacuation travel times and safety areas in a debris flow hazard scenario. Science in the Total Environment. 2020;**712**:136452 [Internet]. Available from: https://www.sciencedirect.com/science/article/pii/S0048969719364484

[9] Golubenko I, Lyamin S, Palymsky B, Goryachev N. GIS of precious metaydeposits of the magadan region: development and creation. Proceedings in International Multidiscipline Science Geology Conference SGEM. 2012;**2**:1073-1077. [Internet]. DOI: 10.0.21.217/sgem2012

[10] Oliveira SC, Zêzere JL, Catalão J, Nico G. The contribution of PSInSAR interferometry to landslide hazard in weak rock-dominated areas. Landslides. 2015;**12**:703-719. [Internet]. DOI: 10.1007/s10346-014-0522-9

[11] Santos T, Deus R, Rocha J, Tenedório JA. Assessing sustainable urban development trends in a dynamic tourist coastal area using 3D spatial indicators. Energies. 2021;**14**(16):5044. [Internet]. DOI: 10.3390/en14165044. Available from: https://www.scopus.com/inward/record.uri?eid=2-s2.0-85113463799&doi=10.3390%2Fen14165044&partnerID=40&md5=4c151ff01bb3f195844a84ec82923f0b

Section 2

Health and Environment

Chapter 2

The Socio-Economic Factors of the Covid-19 Pandemic in Turkey: A Spatial Perspective

Sevgi Eda Tuzcu and Esra Satıcı

Abstract

This study investigates the role of various socioeconomic determinants and vaccination rates in the spread of Covid-19 in a spatial setting in Turkey. For this aim, we employ the 41 sub-indicators of Life Index in Provinces data provided by the Turkish Statistical Institute which is obtained based on the Organization for Economic Cooperation and Development (OECD) Better Life Index approach. Our results indicate no global interactions in the transmission process of the disease among Turkish provinces. This means that the infection burden in the neighboring province does not significantly affect the infection burden of a given state. Yet, we show that vaccination rates and the median age of a neighboring province significantly affect the number of total cases in a given province. We find that as the vaccination rates of a neighboring province rise, the number of total cases in a given province also increases. This finding can be attributed to the "*neighbor–reliant immunity*" concept. It seems that people with vaccine hesitancy toward Covid-19 feel safer without a vaccine when their neighbors are mostly vaccinated. Last, people with a higher satisfaction rate with their health status are more likely to catch the disease due to underestimation of negative consequences.

Keywords: spatial regression, SLX model, Covid-19 cases, vaccination rates, socioeconomic factors, Turkey

1. Introduction

Coronavirus Disease 19 (Covid-19) marked the years 2020 and 2021 with its very fast diffusion rates and severity. With the quick development of vaccines against the disease, the pandemic right now seems to come to an end. Yet, living the last 2 years with a contagious disease has left some serious questions: What is the role of socio–economic determinants in the transmission of an airborne contagious disease like Covid–19? What factors are most influential and make countries more vulnerable to such diseases? What is the role of spatiality in the spread? In this study, we aim to investigate the answers to these questions for Turkey. More specifically, we try to point out the most influential socio–economic factors in the spread of Covid-19 in Turkey in a spatial setting.

The first Covid-19 case is confirmed in Turkey on 11th March 2020 in İstanbul. It spread quickly all over the country. To limit its transmission among the Turkish provinces, similar strategies to other countries, such as travel restrictions and partial curfews, were applied in the initial days. Yet, in time, it has become clear that every country has its own dynamics that limit the effectiveness of precautions against the Covid-19. For example, [1] find that the extreme poverty level is an important determinant in the national performance of low– and middle–income countries, since it determines the ability of social distancing. They also note that the disadvantaged share of the population in terms of socio–economic status is more vulnerable to contagious diseases. Therefore, each country must be assessed individually to understand its needs and to be prepared for future diseases. Analyzing the spread of the Covid-19 and the socio–economic determinants behind is important to be ready for any country as well as Turkey.

The ties between the socio–economic status in the spread of Covid-19 were discussed previously in the literature. These studies mainly focus on mainland China [2, 3] and the USA [4, 5]. Some of them compare the national performances of many countries based on the socio–economic variables, (e.g., [6–8]). Yet, as [4] clearly state, "*In a quickly changing pandemic landscape...county-level data and analysis is crucial to understanding needs and supporting planning efforts.*" We, therefore, turn our attention to Turkey, which is one of the most affected countries in the world. Jain and Singh [9] indicate that 60% of the cases in Asia clustered in Turkey alongside mainland China and Iran. Yet, the number of studies examining the impacts of these variables on the spread of Covid-19 is still limited (among these studies, one can note the study by [10]. Our paper aims to fill this gap while considering the effects of being close to the places where the Covid-19 cases are dense.

Ref. [11] emphasize the role of spatiality in the analysis of contagious diseases by stating that "*when people move, they take contagious diseases with them*.". Much before the Covid-19 pandemic, [12] indicates that infectious diseases are the main concerns of medical geography which defines the "place" as a vital dimension of the transmission process besides the other risk factors. In the SARS epidemic example, [13] notes the importance of detecting spatial linkages which shows the potential spreading ways and spatial clusters. Similarly, [14] argues that the diffusion of infectious diseases is directly related to the location. As a result, to understand the diffusion process of such diseases, spatial analysis is a requirement.

Although the importance of location in the transmission process of such diseases besides the other risk factors is mentioned heavily in the literature, studies considering geography in the Covid-19 incidence rates are scarce and they mostly make a choice between the spatial autoregressive model (SAR) and spatial error model (SEM). Ehlert [15], for example, attempts to determine the socio-economic and region-specific in the Covid-19 transmission in German counties with a choice between SAR and SEM specifications. Andersen et al. [16] examine the local transmission of Covid-19 cases in the USA. Again, they made a selection between SAR and SEM based on the Lagrange Multiplier (LM) tests. Sun et al. [17] employ SAR, SEM, and SAC models to detect the Covid-19 period prevalence in the US counties. Baum and Henry [4] consider several demographic factors and income as well as air pollution and health-related variables in order to explain the spread of Covid-19 in the US states. They also employ a SAR model. Guliyev [18] use the number of confirmed new cases in mainland China as the dependent variable where the recovered cases and the rate of deaths are the explanatory variables in a spatial panel setting. He compares SEM and SAR models, but cannot show spatiality in the explanation of the rate of new cases. He concludes that the spatial lag of X (SLX) model fits the nature of local spillovers in this association for China.

The situation for the scarce studies that consider the spread of Covid-19 in Turkey from a spatial perspective is parallel to the world literature. Tuzcu [8] provides an exploratory spatial analysis with different weight matrices for Turkish Covid-19 cases and deaths in which high spatial autocorrelation is detected particularly for major Turkish provinces. Similarly, [19] use Moran I and Local Indicator Spatial Association (LISA) statistics to determine the hot and cold spots among Turkish provinces. Dinç and Erilli [20] examine the effects of a group of socio–economic determinants as well as climate–based variables on the number of Covid-19 cases with SEM and SAR specifications. Göktaş [21] looks at the relationship between centrality in terms of trade, transportation, and health and the number of cases in a Turkish province while considering other socioeconomic factors as control variables. For this aim, he employs SAR and SAC models. Aral and Bakır [22] use the impact of population density, elderly dependency ratio, Gross Domestic Product (GDP) per capita, literacy rate, and health capacity variables to explain the diffusion of Covid-19 in Turkey with a SAR model. They find global spillovers and significant coefficients for population density and elderly dependency ratio while explaining the increase in the Covid-19 cases.

With this study, we also contribute to the scarce literature on Covid-19 studies in Turkey with a spatial perspective. One of the novelties of this paper comes from the spatial model it adopts. Unlike the previous spatial studies on Covid-19 diffusion, we argue that a spatial Durbin model (SDM) must be the first model to adopt for the analysis. The SDM approach is well known for containing both the global and local spillovers at the same time, which is a feature of the Covid-19 pandemic. In fact, when the best describing model is unknown, [23] suggests using SDM as a starting point as well. As a result, we start our analysis with an SDM setting to detect the local and global spillovers in the diffusion of Covid-19 cases across 81 Turkish provinces. Different from the existing studies, we use the vaccination rates and sub–indicators of Life Index in Provinces by the Turkish Statistical Institute (TSI) as the explanatory variables. Life Index in Provinces report includes 41 sub–indicators about income, work life, safety, housing, environment, social life, access to infrastructure services, education, life satisfaction, and civic engagement. By using these sub–indicators, we believe that every aspect of socioeconomic status in Turkish provinces, from per km2 green area to health capacity, can be taken into account. Hence, an exhaustive list of variables that have the potential to impact the spread of Covid-19 is considered. Controlling the vaccination rates also allows us to detect its role among other variables and its impact on the spread of the disease. By doing so, we are able to contribute to the very limited literature on Covid-19 vaccine hesitancy.

To the extent of our knowledge, a similar study to our setting that examines the spread of Covid-19 in Turkey belongs to [24]. He employs 11 leading indicators of the Life Index in Provinces report, not the sub–indicators as well as other socioeconomic and environmental variables such as GDP, household size, age, air quality, humidity, and average temperature. Although this study also mentions the spatial distribution of Covid-19 cases in Turkey and benefits from some spatial maps, the main analysis method is Ordinary Least–Squares (OLS), not spatial models. By using spatial analysis methods with an exhaustive set of socioeconomic indicators, we believe that our study closes an important gap in the literature.

Our results indicate no significant global impacts in the spread of Covid-19 cases across Turkey, but significant local interactions. We show that vaccination in a given province decreases the total number of cases per hundred thousand people in the same province, but increases the Covid-19 cases in the neighboring province. This seemingly puzzling finding is a result of vaccine hesitancy toward Covid-19 vaccines. The "*neighbor–reliant immunity*" argument by [1] explains that people with vaccine hesitancy feel safer when more people around are vaccinated, so they can act more freely. This situation significantly and negatively affects the total number of cases. We also find that people that are more satisfied with their health status act more carelessly, and the number of total cases increases significantly with higher levels of this variable. The median age of neighbors and the satisfaction rate with a social life are variables that are inversely related to the number of total cases. As the median age of neighbors increases, the social interactions and traveling between provinces decreases to avoid the negative consequences of Covid-19. On the contrary, the rate of membership to political parties in a given province is positively related to the total number of cases in the same province. This finding can be attributed to more social interactions and less social distancing with increased civic engagement.

Based on the findings of this study, we can suggest that the usage of clear communication channels with society has vital importance in fighting against infectious diseases. In this way, it is possible to correct the misperceptions both about the nature of the disease and the vaccinations. Overconfidence about the health status and vaccine hesitancy might increase the overall number of cases, so the burden on the health care system.

The rest of the study continues with an explanation of the data and methodology utilized. The next section presents our findings. The last section concludes with the policy suggestions to the Turkish authorities for the next pandemics.

2. Data and methodology

2.1 Dependent variable and the selection of socio-economic variables

This study uses the total number of confirmed COVID-19 cases per 100,000 population from 81 Turkish provinces as the dependent variable. This data is publicly available and reported as weekly averages by the Turkish Ministry of Health.

To be able to determine which explanatory variables might be important in the spread of Covid-19, we examine thoroughly the previous literature that applies both a spatial and non-spatial analysis. Bassino and Ladmiral [25] argue economic variables like wealth or income are the main drivers of the person-to-person spread of infectious diseases such as COVID-19. Bassino and Ladmiral [26] demonstrate that low literacy has been influential in the spread of the disease. Sun et al. [17, 27] find that age is effective on the spread of COVID-19 cases. Population and population density were also noted as significant variables by [15, 25, 28, 29]. The number of doctors and the number of hospital beds are considered important factors by [3, 25, 26] because their availability has the potential to draw more COVID-19 patients to the area. Living in an urban vs. rural area might be another determinant in the spread of cases as noted by [25, 29]. Ehlert [15] also considers household size as a factor. Social life indicators and average space available per household are used by [25].

We proxy the socioeconomic and health status of each province that is noted in the literature by using the Life Index in Provinces provided by the Turkish Statistical Institute in 2016. This index is produced based on the approach of the OECD Better Life Index. The aim of the Life Index in Provinces is to compare the well-being and living quality of Turkish provinces as well as their economic status. To do so, 11 leading indicators and 41 sub-indicators that include both objective and subjective

DOI: http://dx.doi.org/10.5772/intechopen.106048

aspects of life are created. These indicators include income, work life, safety, housing, environment, social life, access to infrastructure services, education, life satisfaction, and civic engagement dimensions. Based on the previous literature, we select explanatory variables among the 41 sub–indicators. The dimensions that can affect the spread of Covid-19 but are not captured by the Life Index in Provinces, such as median age, percentage of individuals 65 years old and above, or population density are also added to the analysis.

Besides the socio–economic factors, the vaccine uptake decision of societies is a crucial weapon against the spread of Covid-19. Therefore, we use the vaccination rates for individuals 18 years old and above for each province as a control variable in the models.

Variable	Proxy	Source
COVID-19 variables	Total number of cases per 100,000 population by Turkish provinces	Turkish Ministry of Health (weekly)
	Percentage of 18+ population vaccinated against COVID-19 at least once by Turkish provinces	Turkish Ministry of Health (daily)
Housing Conditions	Number of rooms per person	Life Index in Provinces by Turkish Statistical Institute (2016)
	The household size in Turkish provinces	Turkish Statistical Institute (2020)
Work Life	Employment Rate	Life Index in Provinces by Turkish Statistical Institute (2016)
	Unemployment Rate	Life Index in Provinces by Turkish Statistical Institute (2016)
	Average Daily Earnings	Life Index in Provinces by Turkish Statistical Institute (2016)
Income and Wealth	Percentage of Households in middle and higher Income Groups	Life Index in Provinces by Turkish Statistical Institute (2016)
	Percentage of Households declaring to fail on meeting basic needs	Life Index in Provinces by Turkish Statistical Institute (2016)
	GDP per capita by Turkish provinces	Turkish Statistical Institute (2013)
Health	Infant Mortality Rate	Life Index in Provinces by Turkish Statistical Institute (2016)
	Life Expectancy at Birth	Life Index in Provinces by Turkish Statistical Institute (2016)
	Satisfaction Rate with Health Status	Life Index in Provinces by Turkish Statistical Institute (2016)
	Health Capacity Index	Turkish Ministry of Health, Health Statistics (2018)
Education	Percentage of higher education graduates	Life Index in Provinces by Turkish Statistical Institute (2016)
Safety	Murder Rate	Life Index in Provinces by Turkish Statistical Institute (2016)
	Percentage of people feeling safe when walking alone at night	Life Index in Provinces by Turkish Statistical Institute (2016)
Civic engagement	Voter turnout at local administrations	Life Index in Provinces by Turkish Statistical Institute (2016)

Variable	Proxy	Source
	Rate of membership to political parties	Life Index in Provinces by Turkish Statistical Institute (2016)
	Percentage of persons interested in union/ association activities	Life Index in Provinces by Turkish Statistical Institute (2016)
Access to Infrastructure Services	Number of internet subscriptions (per hundred persons)	Life Index in Provinces by Turkish Statistical Institute (2016)
Social life	Number of cinema and theater audiences (per hundred persons)	Life Index in Provinces by Turkish Statistical Institute (2016)
	Shopping mall area per thousand people (m2)	Life Index in Provinces by Turkish Statistical Institute (2016)
	Satisfaction rate with social relations	Life Index in Provinces by Turkish Statistical Institute (2016)
Median Age	Median of individuals' age in Turkish provinces	Turkish Statistical Institute (2020)
Age 65+	Percentage of population over 65 + by Turkish provinces	Turkish Statistical Institute (2020)
Population	Population density of Turkish provinces	Turkish Statistical Institute (2019)

Table 1.
Socioeconomic and Covid–19 related variables and the data sources.

A summary of explanatory variables that are employed in this analysis and the data sources are reported in **Table 1**.

The data period is determined by the announcement periods of the Turkish Ministry of Health. Vaccination rates started to be announced at the province level on 04.07.2021 on a daily basis. The total number of cases per 100,000 population is announced weekly. Therefore, we consider the average total number of cases and vaccination rates for July 2021 in this analysis.

2.2 Methodology and model selection process

A standard OLS model is often estimated as a reference for the following spatial models. This study employs the same starting point. To understand the effect of location on the Covid-19 cases, many studies employ SAR and SEM specifications. You et al. [2] note that the SAR model will show how the infection burden in a location is affected by the infection burden in the neighboring locations. SEM is used to understand whether the OLS residuals are correlated to residuals of the neighboring locations. In the lines of [2, 3] also consider a SAC model. They argue that since the SAC model contains a spatial lag and a spatial error term, it can be seen as a combination of these two.

In fact, the spatial model family has a large set of approaches[1], and model selection is a crucial part of its applications. Baum and Henry [4] argues that this selection must be based on the spillover type that the economic theory points out. Unlike [2]'s suggestion, [4] stresses that the SAC model is not the linear combination of SAR and SEM approaches. Not considering the spillover types in the selection of appropriate spatial models leads us to the identification problem noted in [5].

[1] For a detailed discussion, see [8, 24].

Jamison et al. [6] state that the locations that are closer to the center of the pandemic are affected more quickly than the distant ones. However, besides geographical proximity, Covid-19 can spread easily when the locations are connected on a network, such as traveling. It means that both global and local spillovers exist in the diffusion of infectious diseases. We argue in this paper that this nature of Covid-19 can be best captured with an SDM approach. Aydin and Yurdakul [7] also recommends using SDM as a departure point, when the true data generating process, as in the case of Covid-19, is unknown. SDM will also give the linear combination of SAR and SEM specifications [4], as intended by [2, 3].

The OLS model that is used as a benchmark is presented in Eq. (1).

$$y_i = \beta_0 + \beta X_i + \varepsilon_i \tag{1}$$

where y_i is the total number of Covid-19 per 100,000 people in a given Turkish province. β_0 reflects the intercept term and β is the vector of coefficients for the explanatory variables. X_i is the socioeconomic variables that are shown in **Table 1** and ε_i is the error term with iid. We check the OLS assumptions. No multicollinearity problem is detected. The insignificant variables ($p < 0.10$) are excluded from the model in order to refine.

The SDM specification that is used in this paper is shown in Eq. (2).

$$y_i = \gamma_0 + \rho W y_i + \gamma X_i + W X_i \theta + u \tag{2}$$

In Eq. (2), the dependent and the explanatory variables are the same as the OLS model defined in Eq. (1). However, here, we scale both the dependent and explanatory variables with a spatial weight matrix (W). The coefficient ρ reflects the global interactions in the spread of Covid-19 in Turkey, while θ demonstrates the local interactions. u is the error term.

Our model selection process follows [7] and we also compare our results with SAR and SLX specifications. The SAR and SLX models are shown in Eq. (3) and Eq. (4) respectively.

$$y_i = \alpha_0 + \rho W y_i + \alpha X_i + \lambda \tag{3}$$

$$y_i = \delta_0 + \delta X_i + W X_i \theta + \tau \tag{4}$$

The spatial weight matrix used throughout all these models is the same. The elements of W take the value of 1 if two Turkish provinces are neighbors, and zero otherwise.

3. Findings

3.1 Spatial map of Total number of cases in Turkey

First, we examine the spatial variation of the total number of cases in Turkey. The spread of Covid-19 cases across Turkish provinces is shown in **Figure 1**.

The map in **Figure 1** demonstrates that there are regional variations in the diffusion of Covid-19 cases. The total number of confirmed cases increases from west to east of Turkey. We also consider the four main regions of Turkey and statistically compare the average cases in these regions. These regions are defined as follows: *i*. Marmara, Aegean and Mediterranean Regions, *ii*. Black Sea Region, *iii*. Central

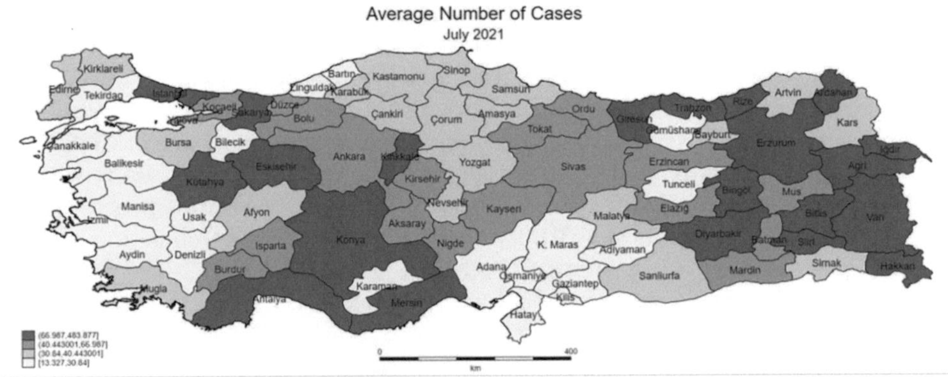

Figure 1.
The variation of Covid-19 cases across provinces in Turkey in July 2021.

Anatolia Region, and *iv*. Eastern and Southeastern Anatolia. Since the normal distribution assumption of ANOVA cannot be satisfied, we compare the means of these regions with the aid of Kruskal–Wallis test. The results reject the equality of means of the Covid-19 cases across regions at a 5% level (χ^2 = 8.757, p–value = 0.0327). Pairwise comparisons revealed that Eastern and Southeastern Anatolia have statistically higher rates than all other regions in Turkey. This result also supports the visual findings in **Figure 1**.

3.2 Results from spatial modeling

We begin our analysis with the classical OLS model. By excluding the insignificant variables at the 10% level, we refine the model and obtain the ultimate model. We check the OLS assumptions. We find heteroskedasticity in our main model which might be a result of spatial dependency.

As explained before, since the association between the total number of cases and the various socioeconomic variables is not well discussed in the previous literature, we start with an SDM specification to avoid the omitted variable problem [23], which is also the linear combination of SAR and SEM specifications [30]. However, the SDM specification does not show significant results. The LR tests comparing SDM vs. OLS and SAR vs. OLS cannot reject the null hypothesis of no significant global interactions. The lack of significant global spillovers indicates that the burden of the disease at one location is not affected by the burden of the disease in the neighboring locations. Yet, the LR test for the coefficients of local interactions in the SDM specification is significant at the 1% level (LR test is 52.5983, and the p–value is 0.0015). That is to say, although no global impacts can be detected in the transmission process of Covid-19 cases in Turkey, geography still matters in the form of local interactions. The socioeconomic features of neighboring provinces are influential on the spread of Covid-19 in a given province. This finding is in line with the study by [18] in which an SLX model is found appropriate to model the new cases in mainland China. Therefore, following [23], we continue our analysis with an SLX model. The final SLX model and the OLS model as a benchmark are shown in **Table 2**.

Vaccination is clearly the strongest weapon in the fight against Covid-19. The findings from **Table 2** also confirm this situation and reveal that the vaccination rate and the total number of cases are significantly and negatively related. Interestingly, it

	(1)	(2)
	Final OLS Model	Final SLX Model
Vaccine Rate	−1.5737***	−1.3693**
	(0.5605)	(0.5926)
Membership to Political Parties	2.0558	3.2805**
	(1.5622)	(1.5043)
satisfaction rate with social relations	−2.1859	−2.2772***
	(1.6857)	(0.7206)
satisfaction rate with health status	3.4398	3.3891**
	(2.6869)	(1.6631)
Constant	−21.0895	−26.1843
	(117.8685)	(105.4470)
W*Vaccine Rate		4.9654***
		(1.7096)
W* Median Age		−10.2514***
		(3.4318)
AIC	870.386	874.219
Adjusted R^2	0.2341	
Number of Observations	81	81

Table 2.
The impact of socioeconomic variables on the total cases of Covid-19: OLS and SLX models.

is found that the effect of vaccination rates in the neighboring provinces is positive and significant. That is, the increased rates of vaccination in the neighboring locations cause a growth in the total number of cases in a given province. This result seems puzzling at first, but it can be explained by the vaccine hesitancy concept. Vaccine hesitancy is defined as the "delay in acceptance or refusal of vaccination despite the availability of vaccination services" [31]. Ke and Zhou [32] state that the vaccine uptake decision of an individual can be dependent on the actions of the neighbors. They call this concept *"neighbor–reliant immunity"*. They argue that people that are hesitant toward the Covid-19 vaccine might feel more "immune" without uptaking the vaccine itself if the people around are already vaccinated. This situation is visible here as well. It is seen that people with Covid-19 vaccine hesitancy do not limit their actions as much as before with neighboring provinces as the vaccination rate of neighboring provinces increases. As a result, the number of confirmed cases in a given province increases.

We also show that as the satisfaction rate with health status increases, the number of total cases also rises in a particular province. This finding can be attributed to the fact that Covid-19 is mostly perceived as an older people's disease or only dangerous for people with co–morbidities. To fight this perception, World Health Organization (WHO) made many announcements, including the one that the Chief of WHO explained that *"young people are not invincible"*. It seems that this perception is still valid in July 2021 in Turkey, and it might grow even stronger with the relatively less severe variants and the ongoing vaccination process.

The rate of membership to political parties is an indicator of civic engagement. As this variable has a higher rate, the social relations, and connections increase as well. This would make it difficult to keep the social distance and adapt to "stay at home" calls. Our findings in **Table 2** confirm this result and demonstrate a positive effect of this variable on the total number of cases.

Median age, itself, is not a determinant of the spread of Covid-19 cases across Turkish provinces. However, the median age of the neighbors negatively impacts the number of cases in a given province. This finding is in line with [15]. He notes that the increase in the median age of neighbors reduces the social interactions with the given state and traveling, so less spread has occurred.

The satisfaction rate with social relations is a proxy for social life. Our results indicate that the higher values of this variable are related to a lower level of total cases. It seems that people who are more satisfied with their social life are most likely to keep their social distance and less engaged with many people. This finding might be explained by the existence of video–calls and other telecommunication methods. Individuals may meet their social needs via the internet and stay at home at the same time.

We cannot show any significant effect of housing conditions, work life, income and wealth, or health indicators other than health status, education, safety, and access to infrastructure services, however.

The results of this paper once more emphasize the importance of vaccinations in order to control the number of cases. In the case of such infectious diseases, governments must use clear communication channels with society to avoid misperceptions about the nature of the disease or the precautions to avoid further spread. Our findings show that over–confidence about the individual health status and vaccine hesitancy increase the number of total cases, so the burden on the health care system.

4. Conclusion

This study employs an exhaustive set of socioeconomic variables and vaccination rates to detect their roles in the spread of Covid-19 in Turkey in a spatial setting. Province-level data allows us to detect the existence of spatiality as well. We cannot detect any global interactions in the diffusion process, so the number of infected people at one location does not bring an extra infection burden to the neighboring locations. Yet, our findings show that local interactions in terms of vaccination rates and median age play an important role in the increase in the total number of cases. Increased vaccination rates in the neighboring provinces also increase the total number of cases in a given province. This result can be explained by the vaccine hesitancy toward the Covid-19 vaccine. We also find evidence that people that are more satisfied with their health status are more likely to catch the disease and increase the total number of cases. To fight the misperceptions about the nature of the disease and the vaccination procedure, the Turkish government must adopt a clear–communication policy and actively work for individuals to access reliable information.

Author details

Sevgi Eda Tuzcu[1*] and Esra Satıcı[2]

1 Ankara University, Ankara, Turkey

2 Independent Researcher, Ankara, Turkey

*Address all correspondence to: stuzcu@politics.ankara.edu.tr

References

[1] Türkoğlu SP, Tuzcu SE. Assessing country performances during the COVID-19 pandemic: A standard deviation based range of value method. Operational Research in Engineering Sciences: Theory and Applications. 2021; **4**:59-81

[2] You H, Wu X, Guo X. Distribution of COVID-19 morbidity rate in association with social and economic factors in Wuhan, China: Implications for urban development. International Journal of Environmental Research And Public Health. 2020;**17**:3417

[3] Kang D, Choi H, Kim JH, et al. Spatial epidemic dynamics of the COVID-19 outbreak in China. International Journal of Infectious Diseases. 2020;**94**:96-102

[4] Baum CF, Henry M. Socioeconomic Factors Influencing the Spatial Spread of COVID-19 in the United States. SSRN; 2020. DOI: 10.2139/ssrn.3614877

[5] Mollalo A, Vahedi B, Rivera KM. GIS-based spatial modeling of COVID-19 incidence rate in the continental United States. Science of the Total Environment. 2020;**728**:1-8

[6] Jamison DT, Lau LJ, Wu KB, et al. Country performance against COVID-19: Rankings for 35 countries. BMJ Global Health. 2020;**5**:e003047. DOI: 10.1136/bmjgh-2020-003047

[7] Aydin N, Yurdakul G. Assessing countries' performances against COVID-19 via WSIDEA and machine learning algorithms. Applied Soft Computing. 2020;**97**:106792

[8] Tuzcu SE. A spatial exploratory look to Covid-19 cases and deaths in Turkey. In: Usupbeyli A, editor. Quantitative Approaches to Current Issues in Economics. Ankara: Ekin Kitabevi; 2020. pp. 119-146

[9] Jain V, Singh L. Global spread and socio-economic determinants of Covid-19 pandemic. Seoul Journal of Economics. 2020;**33**, 40 pages. DOI: 10.22904/sje.2020.33.4.004

[10] Arı A, Özdemir HÖ, Kabadurmuş K, et al. Do Socioeconomic Inequalities Increase the Spread of COVID-19 in Turkey? Applied Economics Letters. 2022. DOI: 10.1080/13504851.2022.2082367

[11] Jia JS, Lu X, Yuan Y, et al. Population flow drives spatio-temporal distribution of COVID-19 in China. Nature. 2020; **582**:389-394

[12] Glass GE. Update: Spatial aspects of epidemiology: The interface with medical geography. Epidemiologic Reviews. 2000;**22**:136-139

[13] Meng B, Wang J, Liu J, et al. Understanding the spatial diffusion process of severe acute respiratory syndrome in Beijing. Public Health. 2005;**119**:1080-1087

[14] Sirkeci İ, Yüceşahin MM. Göç ve Koronavirüs: Nüfus Hareketliliği Verileri Üzerinden KOVİD-19 Salgının Analizi. Göç Dergisi. 2020;**7**:9-34

[15] Ehlert A. The socioeconomic determinants of COVID-19: A spatial analysis of German county level data. Socioecon Plann Sci. 2020;**78**:101083. DOI: 2020.06.25.20140459

[16] Andersen LM, Harden SR, Sugg MM, et al. Analyzing the spatial determinants of local Covid-19 transmission in the United States. Science of the Total Environment. 2021;**754**:142396

[17] Sun F, Matthews SA, Yang T-C, et al. A spatial analysis of COVID-19 period prevalence in US counties through June 28, 2020 where geography matters? Annals of Epidemiology. 2020;**52**:54-59.e1. DOI: 10.1016/j.annepidem.2020.07.014

[18] Guliyev H. Determining the spatial effects of COVID-19 using the spatial panel data model. Spatial Statistics. 2020;**38**:100443

[19] Zeren F, Yilanci V. Anaylsing spatial patterns of the Covid-19 outbreak in Turkey. Bingöl Üniversitesi İktisadi ve İdari Bilim Fakültesi Dergisi. 2020;**4**:27-40

[20] Dinç SC, Erilli NA. Spatial analysis of determinants affecting the total number of Covid-19 cases of provinces in Turkey. ПРИКЛАДНАЯ ЭКОНОМЕТРИКА/Applied Econometrics. 2022;**65**:102-116

[21] Göktaş D. Socioeconomic determinants of provincial Covid-19 cases in Turkey. Working Paper. 2021

[22] Aral N, Bakır H. Spatiotemporal analysis of COVID 19 in Turkey. Sustainable Cities and Society. 2022;**76**:1-10

[23] Elhorst JP. Applied spatial econometrics: Raising the Bar. Spatial Economic Analysis. 2010;**5**:9-28

[24] Kırlangıçoğlu C. Investigating the effects of regional characteristics on the spatial distribution of COVID-19 pandemic: A case of Turkey. Arabian Journal of Geosciences. 2022;**15**:401. DOI: 10.1007/s12517-022-09687-y

[25] Bassino J-P, Ladmiral G. Socio-economic factors influencing COVID-19 spread in Japan virus importation and domestic transmission during the first two waves. Working Paper. 2020

[26] Ramírez-Aldana R, Gomez-Verjan JC, Bello-Chavolla OY. Spatial analysis of COVID-19 spread in Iran: Insights into geographical and structural transmission determinants at a province level. PLOS Neglected Tropical Diseases. 2020;**14**(11):e0008875. DOI: 10.1371/journal.pntd.0008875

[27] Gangemi S, Billeci L, Tonacci A. Rich at risk: Socio-economic drivers of COVID-19 pandemic spread. Clinical and Molecular Allergy. 2020;**18**:10-12

[28] Coccia M. Factors determining the diffusion of COVID-19 and suggested strategy to prevent future accelerated viral infectivity similar to COVID. Science of the Total Environment. 2020;**729**:1-20

[29] Garcia de Alcaniz JG, Romero-Lopez J, Martinez RP, et al. What variables can better predict the number of infections and deaths worldwide by SARS-CoV-2? Variation through time. medRxiv. 2020. DOI: 10.1101/2020.06.04.20122176

[30] LeSage JP. What regional scientists need to know about spatial econometrics. Review of Regional Studies. 2014;**44**:13-32

[31] MacDonald NE, Eskola J, Liang X, et al. Vaccine hesitancy: Definition, scope and determinants. Vaccine. 2015;**33**:4161-4164

[32] Ke Y, Zhou J. The impact of vaccination behavior on disease spreading based on complex networks. Working Paper. 2022

Chapter 3

Spatial Analysis of Climate Driver Impacts on Sub-Saharan African Migration Patterns in Tanzania

Feleke Asrat, Brooks C. Pearson, Matthew H. Connolly and Gizachew Tiruneh

Abstract

Environmental problems resulting from climate change have generated negative impacts on climate-sensitive sectors of African economies. Coping with adverse situations, individuals and households adopt several strategies, including rural-urban migration. Previous literature has investigated the use of migration as a coping strategy to environmental factors. However, specific empirical assessment of links between migration and climatic factors with emphasis on spatial perspectives is not well studied. Accordingly, this study focuses on climatic driver influences on migration from statistical and spatial perspectives using logistic regression and Geographic Information Systems (GIS). We combined secondary data sets collected by the World Bank SLMS nationwide household surveys with geo-referenced sub-villages and historical gridded rainfall and temperature data. Results suggest a significant positive relationship between long-run precipitation and migration, while long-run temperature was statistically inconsequential. Results also suggest spatial patterns and climate change drivers are critical in understanding the migration determinants in Tanzania.

Keywords: climate change drivers, GIS, rural to urban migration, sub-saharan Africa, Tanzania

1. Introduction

Environmental problems resulting from climate change have continued to generate negative impacts on climate-sensitive sectors of African economies. During the twentieth century, the continent faced climate change with a rise in temperature of 0.58°C, with some areas warming even faster than others [1]. Some estimates suggest annual mean surface air temperatures in Africa had risen about 1.5 times the predicted average global increase by 2009 [2, 3]. Despite frequent intense precipitation events, African annual average precipitation has decreased, likely bringing extended droughts. In conjunction with increased prevalence of climate change driven extreme precipitation events, some regions will be increasingly susceptible to both droughts and flooding [2, 4]. This alarming pace of climate change [5] has attracted attention from national and international environmental institutions [3, 6]. Adverse effects of climate change on social and ecological systems include drought, decreased agricultural production, flooding, and hurricanes. These effects are likely to be pronounced in developing countries because their economies are overwhelmingly based on rain-fed agriculture [7–9].

To cope with adverse situations such as climate change, individuals and households adopt several strategies including urban-rural migration [10]. Several studies have recognized migration as one of the coping strategies to environmental factors [10, 11]. However, connections between migration and climate change are not well studied from a spatial perspective. Rural communities of east Africa's Tanzania are following diverse strategies to cope with the climate change impacts, and internal migration is among the many options for regional livelihood improvement. However, complex linkages between climatic factors and migration in Tanzania have been less thoroughly investigated, especially from a spatial perspective.

Therefore, this study assesses the relationship between climate drivers and migration in Tanzania by explicitly incorporating spatial analysis. We statistically described migration pattern responses to climate drivers, and empirically assessed climate driver-migration pattern relationships using regression and geospatial techniques. Through these efforts we tested the following hypotheses: spatial patterns of temperature and precipitation will be different for migrant and non-migrant households; and temperature and precipitation influence migrant more than non-migrant households.

2. Previous studies of migration and climate change

Climate-oriented scientists agree that the likely harmful effects of climate change on rural economies have grown stronger over time. According to [12], almost 870 million globally are chronically undernourished amid climate models predicting rising temperature and declining precipitation for most of Sub-Saharan Africa [6, 12]. Many sub-Saharan economies dominantly depend on climate-sensitive agricultural production, and this region's people are already facing starvation with more than 60% living below the poverty line [13].

According to the Intergovernmental Panel on Climate Change, about 150–200 million people will likely be displaced globally by 2050 due to anthropogenic climate changes [14]. Consensus among global scholars suggests climate change induced migration will worsen in the future [15]. Warming temperatures will have severe effects over the interior semi-arid margins of the Sahara and central southern Africa [1]. Similarly, Africa has remained hot and dry, but is warmer than it was 100 years ago [16–18].

Sub-Saharan Africa precipitation patterns are highly influenced by inter-seasonal and inter-annual climate variability including occasional El Niño-Southern Oscillation events [19–21]. Eastern equatorial Africa is heavily affected by extreme meteorology during the short October–November rainy season, while southeastern Africa experiences similar conditions during the main rainy season in November–February [22–24]. Hulme et al. [16] illustrates the nature of rainfall variability for the Sahel, East Africa, and southeastern regions, and suggests these three regions show contrasting variability. The Sahel shows large multi-decadal variability with recent drying, while east Africa demonstrates a relatively steady regime with some evidence of long-term wetting. By contrast southeast Africa tends to exhibit stability, with some noticeable inter-decadal variation [22], indicating higher latitudes are more susceptible to climate forcings.

Tanzanian climate projections show an expected mean annual temperature increase of 1.7–2.5°C by the 2060s, suggesting a stronger increase than the global average [25]. Similarly, mean annual rainfall patterns are also projected to increase across the country, but with a complex seasonal pattern highlighted by increased January and February rainfall, most dramatically in the far south. For northern Tanzania, highest rainfalls are expected for March, April, and May. During June, July, August, and September, precipitation is projected to increase in the very north of the country, while central and southern Tanzania expect declining rainfall [26].

Migration is a key rural livelihood strategy to increase household earning potential [27], reduce income risk, and shield against socioeconomic and environmental shocks [28]. Migration is a family response to income risk where migrants serve as an income insurance policy for their households of origin [28]. Households effectively diversify income sources by allocating labor to areas with a different set of risks from those faced in the source region, thereby building resilience to various livelihood shocks [29].

In their study on Ecuador, [30] show agricultural shocks are key factors in international migration. Munshi et al. [31, 32] have shown strong links between climate change, crop yields, and migration, whereas [33] discovered an opposite relationship suggesting Mexico to U.S. migration decreases as rainfall declines. Ethiopian studies show rural out-migration responds to drought sensitivity [34, 35].

There seems to be lack of consensus on the role of disasters in international migration with some arguing for a positive link [36–39], while others indicate little or no or a negative relationship [40, 41]. For instance, [40] found flooding has modest to insignificant impacts on migration. On the other hand, [41] discovered people did not migrate after the occurrence of disaster in Bangladesh.

Meze-Hausken [35] showed weather anomaly impact on international migration has two channels. First, weather anomalies will lead to lower rural wages, especially if the effect of weather anomalies on agricultural production is strong. Second, lower rural wages will attract more mobile workers to move from the rural areas to cities in search of work. Consequently, more people settle in urban areas and thus increase urbanization. Therefore, weather anomalies are a key determinant of increased urbanization.

The growing literature examining the migration determinants increasingly emphasizes the role of environmental change in in-migration processes. Nawrotzkia and Maryia [42] show there are different ways to measure climatic factors. Heat waves, cold snaps, droughts, and excessive precipitation can be thresholds to construct climate measures. Schlenker and Roberts [43] investigated temperature effects on corn, soybeans, and cotton based on nationwide crop yield data and growing season climate information. Their findings indicate threshold temperatures for each crop with temperatures above the respective thresholds leading to yield reductions.

The literature therefore suggests strong links between climate change and migration. Climate change and migration relationships are relatively well researched from a non-spatial perspective. Spatially explicit empirical research in this context is in its infancy, especially for regions like Africa. Therefore, the present study seeks to address the paucity of empirical studies in the region. Another important knowledge gap is the literature's focus on precipitation or temperature variation, while neglecting the effect they may have jointly on migrant decisions. Therefore, a clear understanding of climate change impacts on migration requires assessment of spatial links between climate change and migration. Further, impacts of temperature and precipitation variations on individual and household migration decisions need to be assessed in the same framework.

3. Study area

The Republic of Tanzania is located south of the equator at 6°00′S 35°00′E bordering the Indian Ocean and eight eastern and southern African countries. The country is bordered on the south by Mozambique, Malawi, and Zambia; on the west by Zaire, Burundi, and Rwanda; on the north by Uganda and Kenya; and on the east by

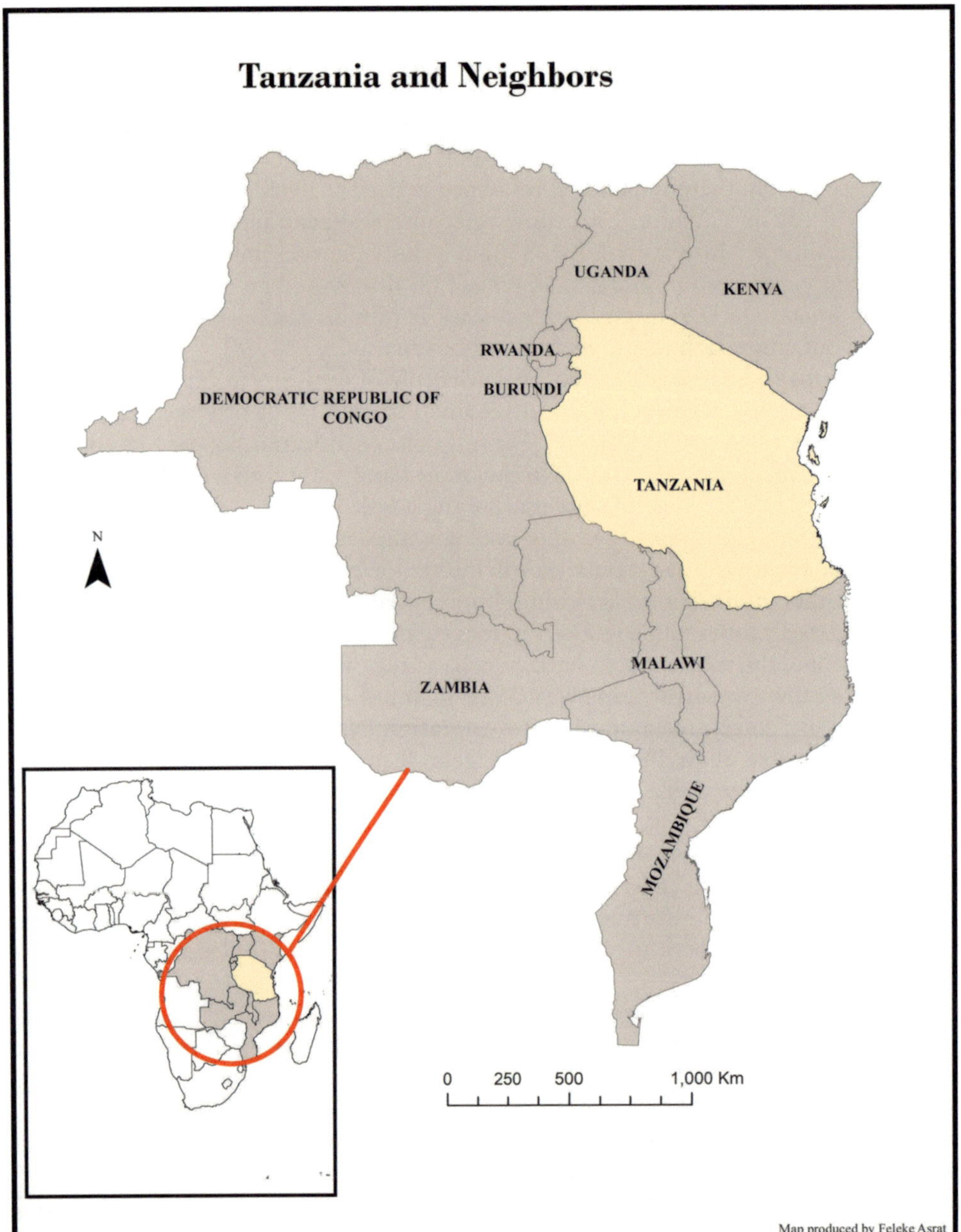

Figure 1.
Study area relative location.

the Indian Ocean (**Figure 1**). The country's total area is 947,300 km^2, with 885,800 km^2 covered by land surface, and 61,500 km^2 of water.

According to the Tanzania Bureau of Statistics [44], the country has a total population of 45 million with a population density of 51 persons/km^2. Tanzania has 30 geographic regions, 25 on the mainland and 5 islands (**Figure 2**). Administratively, the country is divided into regions, districts, wards, and villages.

Agriculture remains the economy's largest sector. According to The World Bank's economic outlook [45], the agricultural sector contributes almost one quarter of

Figure 2.
Tanzanian administrative regions.

Tanzania's overall gross domestic product (GDP), accounting for 85% of country's exports, and employing about 80% of the work force. The majority of the country has a tropical climate with variations across regions. Except for the eastern coastal strip, most of the country is highland or a central plateau approximately 900–1800 m above sea level, with mountain ranges including Mount Kilimanjaro at 5895 m [26].

Tanzania's regions exhibit topographic and seasonal variations in temperature. Accordingly, highland temperatures vary between 10 and 20°C during cold and hot seasons respectively. In the other parts of the country, temperatures rarely fall below 20°C. November through February is the hottest period, ranging between 25 and 31°C, whereas the coldest period occurs between May and August with temperature ranges of 15–20°C [13].

Seasonal Tanzanian rainfall is heavily influenced by Inter-Tropical Convergence Zone (ITCZ) migrations. The ITCZ moves southwards through Tanzania from October to December, reaching the south of the country in January and February, and returning northwards in March, April, and May. This migration brings northern and eastern Tanzania two diverse wet periods with short rains in October to December, and long rains in March to May. Conversely, the southern, western, and central parts of the country experience only one wet season that extends from October through to April or May. Seasonal rainfall varies regionally between 50 and 200 mm per month, with some regions receiving up to 300 mm per month in the wettest season [26].

4. Study data

Our study is based on high-quality household survey data called the Living Standards Measurement Survey-Integrated Survey on Agriculture (LSMS-ISA), initiated by the World Bank Development Economics Research Group, and implemented by the Tanzania Bureau of Statistics [44, 46]. National-level longitudinal data was collected between 2009 and 2013. The present study is specifically based on the 2012–2013 data. The LSMS-ISA data are composed of information on households, agriculture, and community characteristics. Each survey household is associated with a georeferenced sub-village.

Sub-villages represent enumeration areas where households were selected for the survey. However, an enumeration area is not a community from the sociological aspect; instead they are designated for information collection about the study areas where households selected for the intended study are located. A total of 26 regions and 149 sub-villages were considered in this study. Some observations were removed due to incomplete surveys and georeferencing errors, generating a final sample size of 10,461 households. Accordingly, 3968 and 6493 observations were identified for non-migrant and migrant households respectively. Non-migrants were identified in 62 enumeration areas whereas migrants were located in 75. The survey tracked all household members 15 years or older. We focused on these households and individual members aged 15–65 years. A key study variable was identification of migrant and non-migrant individuals. A migrant is an individual in a household who has left his or her initial residence and considers himself or herself to have settled in a new community. Despite the foregoing definition, the study follows the new economics of migration approach as indicated by previous work [47–50], where migration is a collective action made by a household. A household member migrates not only to maximize household income for economic reasons, but also to minimize risk. Therefore, this study analysis used household-level data.

In addition to the LSMS-ISA data, this study incorporated 0.5° gridded historical climate data from the University of East Anglia's Climate Research Unit [51] for 1983–2012. As indicated in the Results section, the 30 year (1983–2012) gridded temperature and precipitation data were downloaded and assigned into sub-villages using *ArcGIS* and *STATA* software. Temperature and precipitation data were analyzed for anomalies as well as for long-run (30 year) temperature and precipitation means.

Variable selection was based on the following studies: tenure security, land size, distance to market [52]; age, number of male and female adults, literacy, tropical livestock unit (TLU) [53]; extension of advice, soil fertility [54]; and climatic variables [11]. Additionally, [54] used social links and irrigation potential as measures of information access and land quality respectively. Similarly, we used extension advice (or access to information) and soil fertility as proxy measures of land quality.

5. Methods

This study examined the ability of 12 independent variables to explain household migration decisions. Two variables are climatic (temperature and precipitation), while four involve household demographics (household head age, household literacy, males over 15, and females over 15). Six variables define household economic characteristics: availability of extension advice, land ownership, TLU, soil quality, total land area, and distance to market.

We analyzed spatial patterns of precipitation and temperature across the regions with *ArcGIS*. The georeferenced datasets were converted into *ArcGIS* shapefiles using the Arc 1960 projection for Tanzania [55]. Households were divided into migrant and non-migrant categories. A household with at least one member who migrated during the survey year (2012) was considered a migrant household. All households in the same sub-village were assigned the same locational coordinates to merge climate data with each household. Temperature and precipitation changes were identified and first standard deviation data were incorporated into the analysis [43]. These data were mapped to show spatial migration patterns for migrant and non-migrant households across the country's regions. Coefficients of variation (CoV), defined as σ/μ, for each sub-village's rainfall and temperatures were displayed and analyzed using choropleth and graduated symbol maps. Average maximum temperature and rainfall represented annual extreme conditions. Long-run average rainfall and temperature represent the mean of yearly average values. Anomalies were the difference between the annual average and the 30 year annual mean. Anomalies are also shown using choropleth and graduated symbol maps. Additionally, household exposure to climatic variables was measured by the between-years rainfall CoV, for the period 1983–2012. CoV provided several advantages. First, for a given level of standard deviation, the CoV changes as the mean changes showing a lower level of variability for sub-villages with higher levels of average rainfall or temperature. Second, CoV shows dispersion of temperature and rainfall values in relative terms, allowing comparison between two sub-villages. We also computed sub-village average precipitation shortfall (i.e., the average of the annual totals' departures from the long-run average). The same procedure was followed for study period temperature.

Migrant and non-migrant household demographic characteristics were analyzed using descriptive and inferential statistics for variables such as age, sex, household size, female head of household, literacy status of household head, male and female workforces, and households' access to agricultural extension. Migrant and non-migrant households; distance from household farm plots to market; total land holding size; and total number of Tropical Livestock Units (TLU) were analyzed with t-statistics.

Our control variables and migration are related as follows. Household educational level is expected to increase migration because better-educated individuals are more likely to have information about migration and job availability in urban areas than are less educated individuals [56]. Total land holding is expected to decrease migration as larger land size corresponds to increased labor requirements for the household. Conversely, larger family size is expected to increase migration since households may have an excess labor supply.

The relationship between household head age and migration is expected to be mixed. If the household head is middle-aged, he may send a family member to migrate since migration may not affect family labor demand. Alternatively, an elderly household head may not be able to perform demanding tasks and may not be able to afford to expend family members on migration.

As a wealth indicator, TLU is expected to have a mixed effect on migration. Poorer households (low TLU) may lack financial resources for migration [57]. However, low income families tend to engage in short distance migration, while better income families may choose long distance migration [58]. Furthermore, gender may be important as female members of the household may not migrate long distances.

A household holding title to land is more secure than a landless household. Increased tenure security may discourage migration. Likewise, households having good soil quality could prefer working their land instead of migration. Household plot distance to market, on the other hand, may positively influence migration since a household close to market may have better labor market information and therefore opt to migrate.

Besides descriptive statistics, t-tests, and spatial analyses, we employed a logistic regression analysis to explain migration. Our analysis was framed following [59] approach, presenting migration as a determinant of a set of explanatory variables. We extended this approach to feature rainfall and temperature as key migration determinants. The logit specification assumes the household chooses migration *m* if the utility derived from the choice of *m* is greater than the decision not to migrate. As the utility from migration/non migration is unobservable, it can be expressed as a function of observables in the latent variable model, given in Eq. (1):

$$M_i^{\,*} = \alpha Z_i + \varepsilon_i \; with \; M_i = \begin{cases} 1 & if M_i^{\,*} \\ 0 & otherwise \end{cases} = 0 \tag{1}$$

where Mi is a dummy variable for the choice of migration; $Mi = 1$ if the household has chosen the decision to migrate, and $Mi = 0$ otherwise. α is a vector of parameters to be estimated; Zi is a vector that represents household characteristics; and εi is the random error term.

Based on Eq. (1), the estimated relationship between migration and temperature and precipitation is given by:

$$\ln\,(m_h) = \alpha_h + \beta x_h + \tau d_h + \rho i_h + \mu e_h + \varphi j_h + \varepsilon_h \tag{2}$$

where h denotes household for migrant or non-migrant family. Migration is denoted by m_h, x_h represents the temperature for each household h, while d_h represents the precipitation for household h over a 30 year period. The set of socio-economic characteristics are denoted by i_h while e_h and j_h are the set of

	TLU	Soil	Title	Area	Temp	Precip	Dist to market
Good soil quality	0.0165	1					
Has title	0.0142	0.0132	1				
Land area	−0.1986	−0.0604	−0.0449	1			
Average temperature	0.0492	−0.0131	−0.0474	−0.0196	1		
Average precipitation	0.0797	−0.0514	−0.0016	−0.0069	0.1582	1	
Distance to market	−0.019	−0.0397	−0.0068	−0.0735	0.0271	0.1042	1

Table 1.
Key independent variable pearson correlation matrix.

institutional and the physical capital variables respectively. The coefficients β, τ, ρ, μ and φ denote the respective vector of parameter estimates, and ε_h indicates the error term.

The model's dependent variable is set to a 0–1 dummy variable, where 1 represents migrants and 0 represents the non-migrants household. Accordingly, the predicted values for the dependent variables will fall between 0 and 1 interval. These results will show the probability of households deciding to migrate.

As previously indicated, this study's first hypothesis states that temperatures and precipitation show different spatial patterns between migrant and non-migrant households across sub-villages in the region. Inferential t-test statistics compared the effects of temperature and precipitation as well as the other control variables on migrant and non-migrant households. The second hypothesis states temperature and precipitation have more influence on migrant than non-migrant households with differing impacts across space. This hypothesis was tested using regression analysis. Regression coefficients quantified the (positive or negative) impact of a unit change in that variable on the propensity to migrate.

Our control variable decisions were based on [60, 61]. Since climatic variables and control variables are exogenous (unaffected by patterns of migration), a logit equation can be estimated without endogeneity concerns. Additionally, we addressed multicollinearity concerns with an independent variable correlation analysis. None of the pairs of variables were highly correlated (**Table 1**).

6. Results and discussion

Table 2 presents demographic characteristics for non-migrant and migrant households. Average long-run temperature and precipitation for the non-migrant households were 23.59°C and 879.37 mm, respectively. The long-run temperature and precipitation for the migrant households were 23.72°C and 876.38 mm. The average age of household heads was 49 years for non-migrants and 50 for migrants, showing little between-group difference. On average there were two males and two females for non-migrant households and three each for migrant households. Seventy percent of non-migrant household heads and 74% of migrant household heads were literate. Household labor supply averaged 1.5 males and 1.6 females for migrants, and 1.2 males and 1.3 females for non-migrants. The household wealth measure (TLU) varied from 3.9 for migrant households to 2.7 for non-migrants.

	Non-migrant		Migrant		Diff	p-value
Variable	**Mean**	**s**	**Mean**	**s**		
Gender	1.28	0.45	1.2	0.4		
Avg. Temp.	23.6	2.7	23.7	2.9	−0.12	0.03
Avg. Precip.	879.4	371.7	876.4	407.0	3.0	0.71
Age	49.4	16.9	50.1	15.2	−0.78	0.01
Literacy status	1.3	0.5	1.3	0.4	0.03	0.00
Male workers	1.3	1.0	1.5	1.1	−0.25	0.00
Female workers	1.3	0.9	1.6	1.1	−0.25	0.00
TLU	2.7	8.2	3.9	10.0	−0.27	0.00
Land area	6.7	8.5	10.4	20.0	−1.19	0.00
Extension	0.09	0.3	0.1	0.3	−3.64	0.00
Dist. to market	11.8	6.6	10.0	7.6	0.20	0.05
Good soil	0.5	0.5	0.4	0.5	0.02	0.05
Has title	0.1	0.3	0.2	0.4	−0.05	0.00

Table 2.
Summary statistics for selected variables by migration status.

Land-related analysis is also considered for both categories of households. Land ownership (tenure security) impacted a household's migration decision. On average, 11.8% of non-migrant household and 17% of migrant households have title to land, with migrant and non-migrant households owning 3.61 and 2.33 hectares of land respectively. Household information availability was important to migration decisions as 13% of migrant households received agricultural and livestock advice, but only 9% of non-migrant households were well informed.

Bio-physical variables included household plot distance to market and soil quality. Non-migrant household plots average 11.8 km from market, while migrant household plots were only 9.9 km from market. By contrast, non-migrant households typically had better quality plots (46% good soil), compared to 44% good soil for migrant households.

Next, we show spatial patterns of climatic factors as well as distance from household plots to nearest market and land-size distributions by region. **Figure 3** depicts the migration patterns by sub-village. Yellow dots depict sub-villages where non-migrants are dominant, while brown dots represent sub villages where the majority are migrants. While migrants and non-migrants are evenly distributed, there are more non-migrants in the northern part of the country as opposed to the southern. Similarly, southeastern Tanzania is non-migrant dominated compared to its northwestern section which has more migrants.

Figures 4 and **5** show average annual temperature and precipitation by sub-village for the period 1983–2012, and indicate spatial variation in highest average annual temperatures from 26.67 (southwest) to 29.52°C (northwest). Lowest temperatures ranged from 15.18 to 18.25°C as one travels northwards in Tanzania. Average annual precipitation for 1983–2012 is shown in **Figure 4**, which ranges between 1616.23 and 2113.44 mm, which is mainly received in parts of the southern and central sections of the country.

Figures 6 and **7** illustrate long-term temperature and precipitation conditions through Tanzanian space. The coefficient of variation is defined as the long-term deviation divided

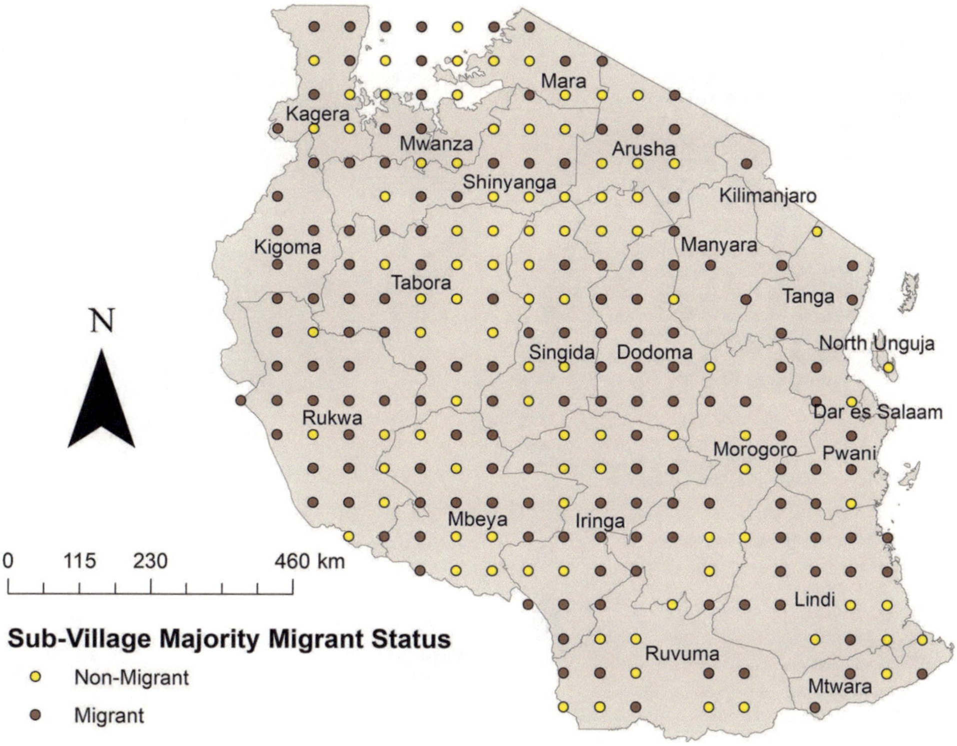

Figure 3.
Tanzanian migration patterns.

by the long-term mean for temperature and precipitation taken independently (Charles et al. 2005). A negative coefficient of variation indicates an area is semi-arid or arid, while a positive coefficient of variation indicates sub-humid and humid environments. **Figure 6** shows regions of low variability were observed in the central and eastern part of the regions for temperature. As shown in **Figure** 7 high precipitation variability areas were dominantly in the northwest, central, and southern regions.

Figure 8 shows average distance from household plots to the nearest market by sub-villages. Shortest distance ranges varied from 0.00 to 5.80 km, with the greatest variability observed in the east. On the other hand, longest observed distances ranged from 81 to 190 km along the northern, central and southern regions. **Figure 9** depicts household land size spatial distributions across the country. In general, plot sizes are smallest in the southeast, while plots are largest in the west and northeast.

In general, the foregoing spatial analyses seem to show the following patterns and correlations between the migration and climatic variables. Long-run average precipitation statistical results show that the coefficient of rainfall has a significant and positive impact on migration. Spatial results indicate south and central Tanzania mainly received the highest rainfall. Migration is significantly more likely from the southern and central sections of the country where rainfall is the highest. Spatial analyses also seem to show that distance of plot to market have a significant negative impact on migration. Moreover, spatial results indicate the shortest distance recorded ranged from 0.00 to 5.80 km, mainly observed in the eastern part of the country. Therefore, migration is significantly less likely from eastern parts of the country where average distance from plots to market is the shortest. Finally, household

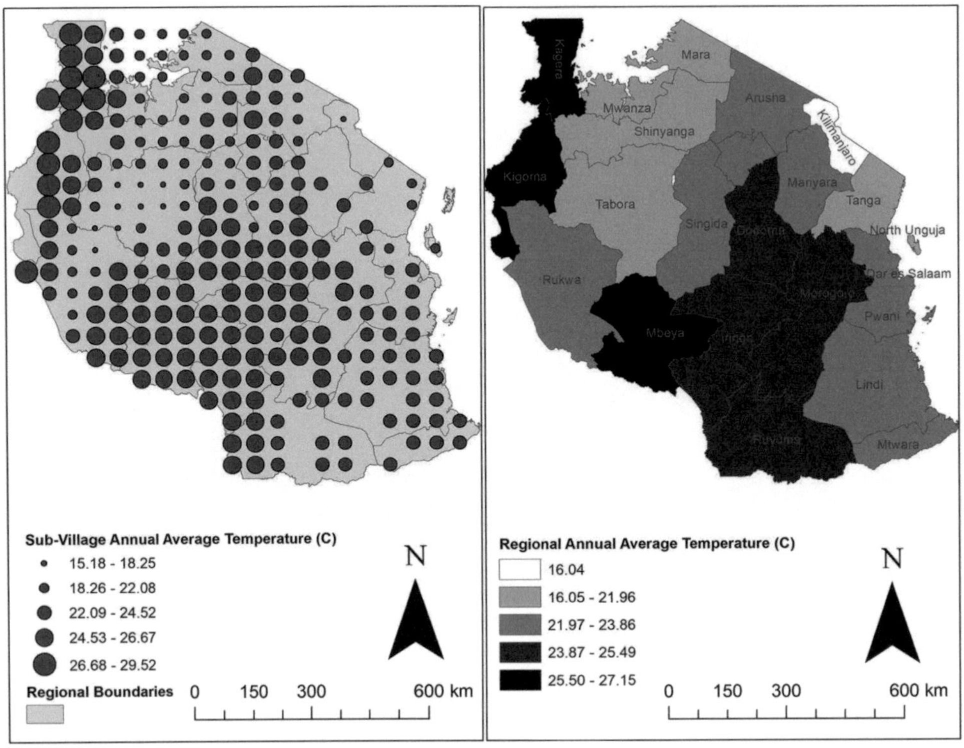

Figure 4.
Long-run temperature by sub-village and region.

land-size tends to be a significant migration determinant. Spatial results indicate that western and northeastern Tanzania have most of the largest farm plots. Thus, the impact of land-size on migration will be stronger in the west and northeast.

In **Table 3**, we show the logistic regression outputs for the determinants of migration, in which the interpretation of the results is based on the log likelihood ratio. Despite a poor fit to the data as indicated by the Hosmer-Lemeshow (HL) likelihood ratio test significance (a pseudo R^2 of 3.6%), all 12 original variables but two were found to be significantly affecting migration.

The empirical analysis indicates climate expressed by long-run precipitation had a significant positive effect on migration, where a 1 millimeter rise in precipitation led to a likelihood of 0.0034 migration increase. By contrast, long-run temperature had an insignificant impact on migration. These findings have two implications. First, households respond to rainfall instead of temperature. Second, rainfall is a potential migration driver. It should be noted that the descriptive statistics discussed above show that there is a significant difference in temperature (and not precipitation) between migrant and non-migrant households' locations. However, the descriptive statistics suggest differences in magnitude and not climatic factors which may impact migration. This observation alone, however, because the t-test is indicative of stock differences between the two variables with respect to migration while the logistic regression shows the impact of each of variables on migration of a given variable when everything else is held constant. There are multiple possible explanations for this apparent paradox in study results including the potential influence of flooding on migration decisions which is beyond the scope of this study.

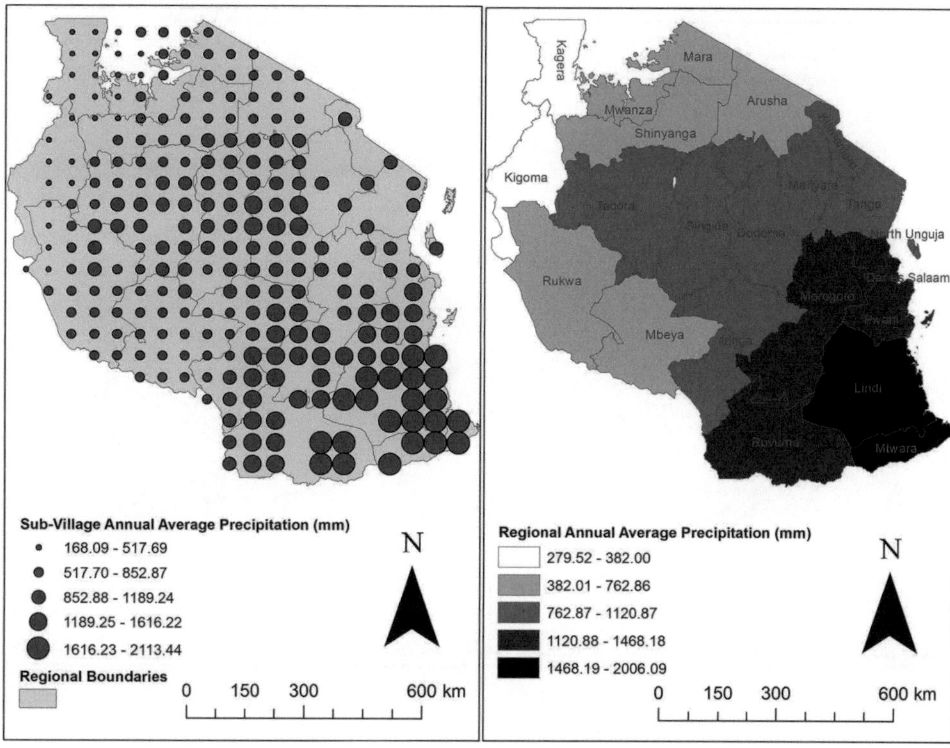

Figure 5.
Long-run precipitation by sub-village and region.

Age of household head, household head literacy status, and male and female workforce variables demonstrate significant robust impacts on migration. Access to agricultural extension services also shows significant and strong effects on migration. This analysis suggests households with older members, family with more male and female workforce, and households with access to agricultural extension services tend to migrate out of their community more so than other households. On the other hand, higher level of literacy seems to discourage migration.

The total livestock unit (TLU) variable showed insignificant impact on migration. Since livestock ownership serves here as a proxy indicator for wealth in rural areas of Tanzania, a higher TLU household is a richer family. However, larger household income seems both to encourage and discourage a given member of a family to migrate. Empirical results suggest households working on good soil quality farms and with farm plots far from market tend not to migrate. Households with good soil conditions have the opportunity to earn more revenue from farm produce than households with poor soil. Thus, the household's tendency to improve livelihoods by sending a household member to migrate would be lower for those with better soil quality. The negative effect of distance to market indicates household access to a nearby market reduces households' tendency to send a family member to migrate. This outcome could occur because households closer to markets do not see the need to migrate because of greater economic opportunity compared to households farther from market [60].

Study results suggest families holding title to their land are 0.0787 times more likely to migrate. Total land area is also significantly related to migration. When

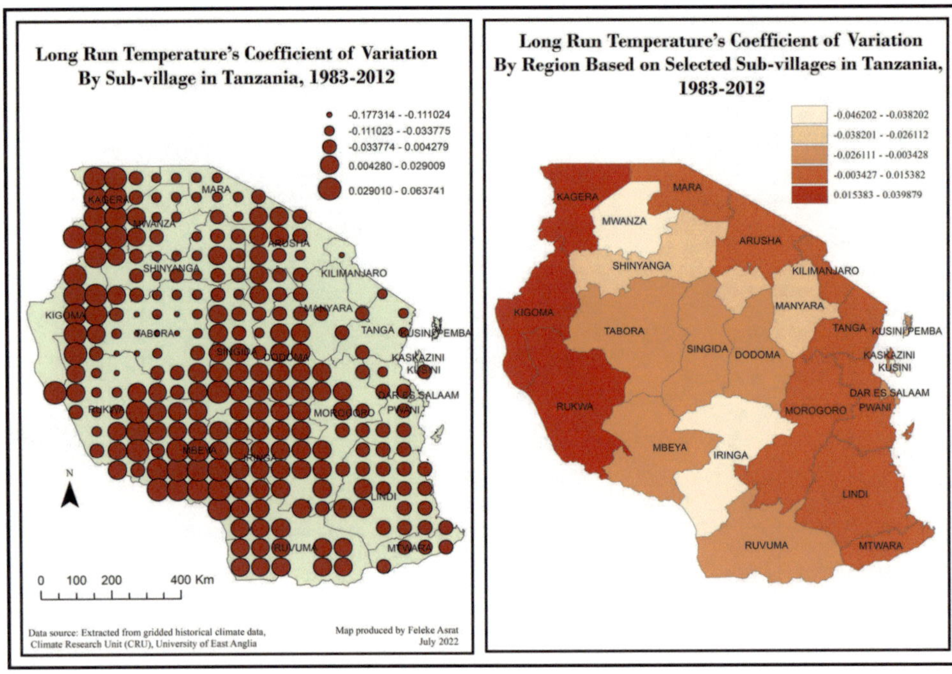

Figure 6.
Long-run temperature coefficient of variation (CoV).

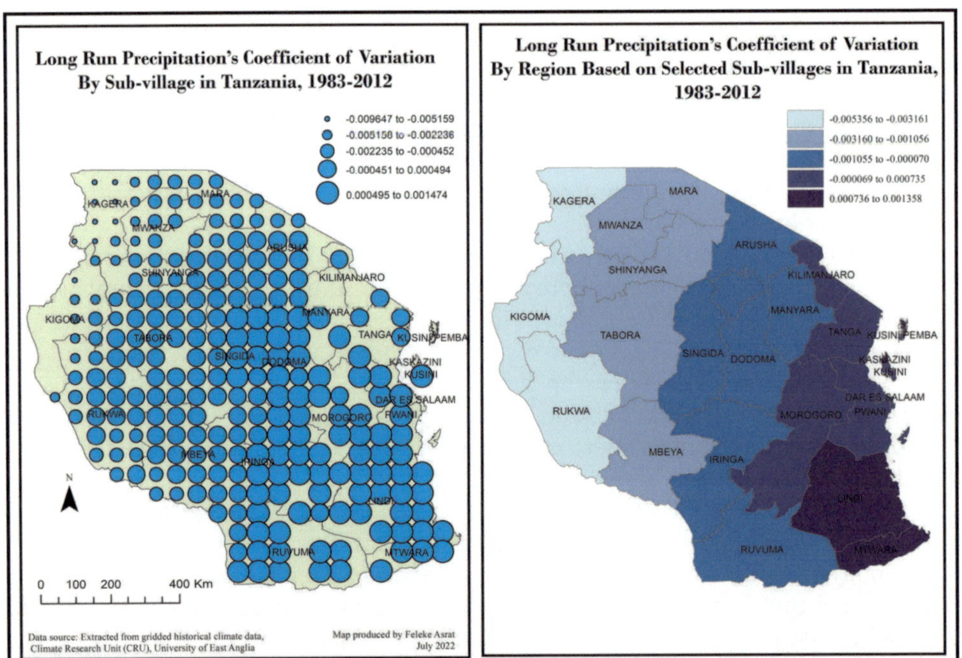

Figure 7.
Long-run precipitation coefficient of variation (CoV).

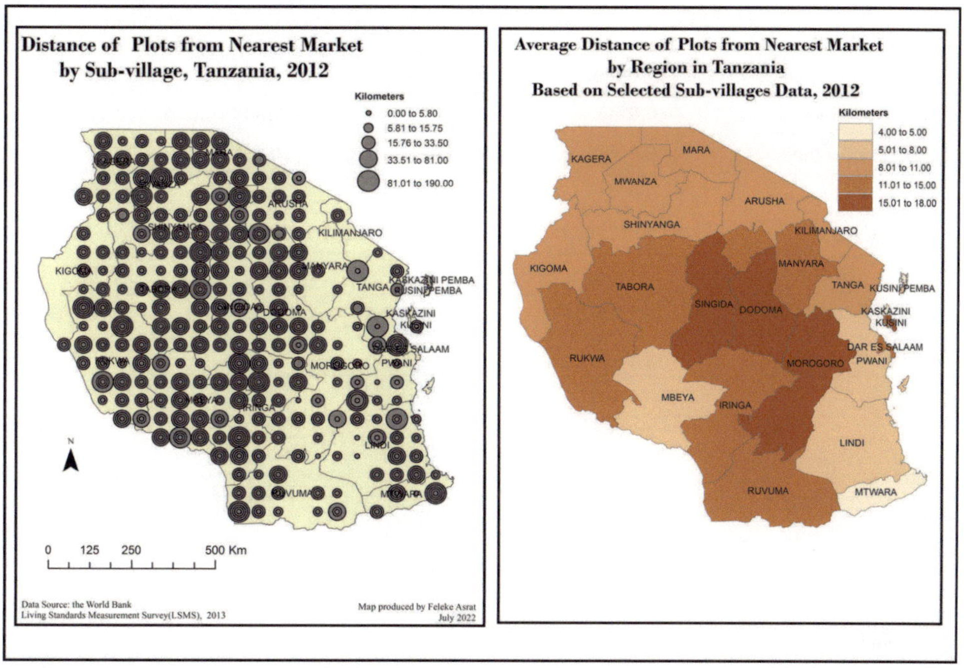

Figure 8.
Household plot distance to nearest market.

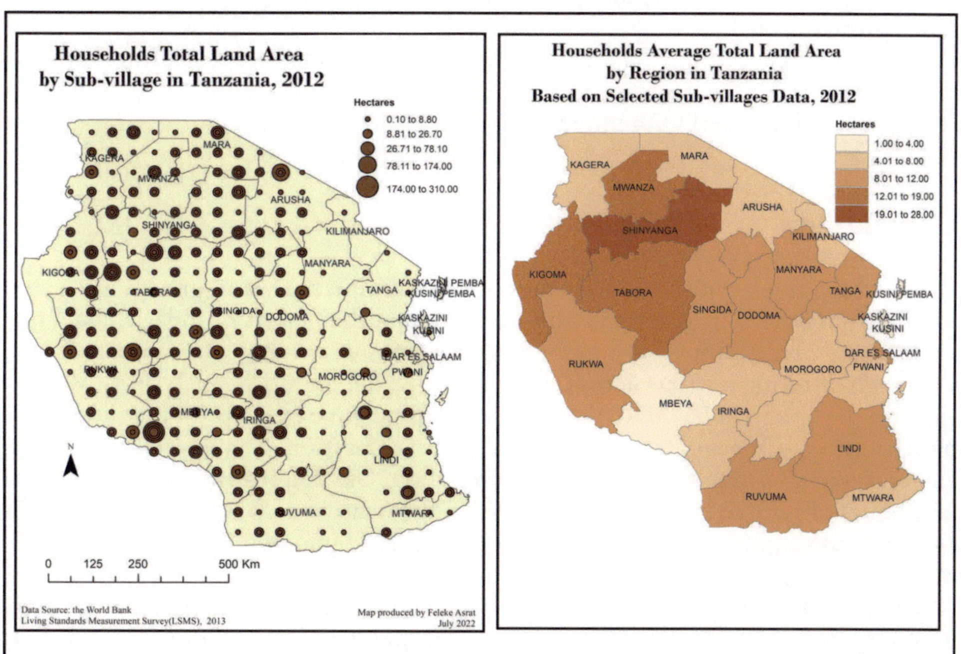

Figure 9.
Household land size.

Variable	Coefficient			
	Estimate	**Std error**	**p-value**	**B(exp)**
Avg. Precipitation	0.0148	0.0075	0.048	0.0034
Avg. Temperature	−0.00003	5.38×10^{-5}	0.632	5.99×10^{-6}
Household head age	0.0041	0.0014	0.002	0.0010
Literacy status	−0.1220	0.0487	0.012	−0.0284
Male workforce	0.1306	0.0218	0.000	0.0304
Female workforce	0.2025	0.0238	0.000	0.0471
Extension advice	0.1920	0.0691	0.005	0.0437
TLU	−0.0017	0.0030	0.572	−0.0004
Soil quality	−0.0945	0.0413	0.023	−0.0220
Has title	0.3511	0.0606	0.000	0.0787
Area	0.0254	0.0027	0.000	0.0059
Distance to market	−0.0110	0.0014	0.000	−0.0025
Constant	−0.4619	0.2126	0.03	
Log likelihood ratio	−6693.4207		HL chi-square	499.66
Pseudo R^2	0.036		HL p-value	0.0000

Table 3.
Logistic regression results for migration determinants.

considering both title of land and total land area, the analyses indicate both variables show significant positive migration impacts. These outcomes suggest households with more land holdings and secured land ownership tend to migrate. Since land serves as financial capital for tenure-secured households, land ownership might be used to leverage costs associated with migration.

7. Conclusions

Our study investigated two hypotheses. First, temperature and precipitation show spatial differences in migrant and non-migrant households in Tanzania. Second, temperature and precipitation have more influence on migrant than non-migrant households with differing impacts across space. Specifically, empirical analysis shows that long-run precipitation has statistically significant positive impacts on migration and suggests increased precipitation is a key driver of migration. Conversely, long-run temperature had insignificant effects on migration. Our findings seem to suggest that the amount of rainfall is more important for Tanzanian households since their livelihood depends on it. On the other hand, some level of temperature variation may be tolerable as long as the rain is falling.

In sum, this study provides insight into spatial climate driver impacts on migration using georeferenced household survey data and gridded historical precipitation and temperature data for Tanzania. The study also contributes knowledge on spatial climate driver impacts on migration by identifying relevant determinant variables for Tanzania. This study makes an important contribution on two fronts. First, it adds to the limited empirical literature on Africa that assesses the impact of climatic factors in a detailed manner. Second, it combines econometric and spatial perspectives in the analysis to quantify key relationships while illuminating spatial pattern differences for key variables.

A shortcoming of our study is the loss of spatial resolution due to sub-village aggregation of household-level survey data. This spatial aggregation greatly limited the spatial analyses' flexibility. Consequently, future studies with better spatial data could provide more accurate results. In addition, the statistical tests indicated overall poor model fitness which could be rectified with additional explanatory variables and well-refined data.

An additional possibility worth exploring is a gender differentiated migration response to climatic and non-climatic shocks because adult men and women have distinct roles in agricultural activities and have different levels of land tenure security [62]. Therefore, investigating gender-based differences in migration would be an interesting addition to the literature.

Author details

Feleke Asrat, Brooks C. Pearson*, Matthew H. Connolly and Gizachew Tiruneh
University of Central Arkansas, USA

*Address all correspondence to: bpearson@uca.edu

References

[1] Eriksen SK et al. Climate Change in Eastern and Southern Africa: Impacts, Vulnerability and Adaptation. U of Oslo and NORAD: Global Environmental Change and Human Development; 2008

[2] Boko M, et al. Africa. Climate change 2007. In: Parry ML, et al. (Eds). Impacts, Adaptation and Vulnerability. Contribution of Working Group II to the fourth assessment report of the Intergovernmental Panel on Climate Change. Cambridge: Cambridge UP; 2007. pp. 433-467

[3] Yanda Z, Mubaya P. Managing a Changing Climate in Africa Local Level Vulnerabilities and Adaptation Experiences. Dar es Salaam: Mkuki na Nyota; 2011

[4] World Wide Fund for Nature (WWF). Climate change impacts on East Africa: Review of Scientific Literature. Gland, Switzerland: WWF-World Wide Fund for Nature; 2006

[5] Mora C et al. The projected timing of climate departure from recent variability. Nature. 2013;**502**:183-195

[6] Intergovernmental Panel on Climate Change (IPCC). Climate Change 2014: Mitigation of Climate Change. Intergovernmental Panel on Climate Change. New York: Cambridge UP; 2014. Available from: http://www.ipcc.ch/report/ar5/wg3/ [Accessed: May 13, 2019]

[7] Cranfield L et al. Model selection when estimating and predicting consumer demands using international cross section data. Empirical Economics. 2003;**28**(2):353-364

[8] Deschenes O, Greenstone M. The economic impacts of climate change: Evidence from agricultural output and random fluctuations. The American Economic Review. 2007;**97**(1):354-355

[9] World Bank. World Development Report 2010: Development and Climate Change. Washington: World Bank; 2010

[10] Barrios SL et al. Climatic change and rural-urban migration: The case of sub-Saharan Africa. Journal of Urban Economics. 2006;**60**:357-371

[11] Hatton J, Williamson G. Demographic and economic pressure on emigration out of Africa. Scandinavian Journal of Economics. 2003;**105**(3): 465-486

[12] Food and Agriculture Organization (FAO), World Food Program (WFP), International Fund for Agricultural Development (IFAD). The State of Food Insecurity in the World 2012. Rome: FAO; 2012

[13] United Nations Development Program (UNDP). Human development report 2007/2008: Fighting Climate Change: Human Solidarity in a Divided World. New York: UNDP; 2007

[14] Stern N. The Stern Review: The Economics of Climate Change. Cambridge: Cambridge UP; 2006

[15] Warner K et al. In Search of Shelter: Mapping the Effects of Climate Change on Human Migration and Displacement. UN University, CARE International, Columbia University, UNHCR and The World Bank; 2009

[16] Hulme M et al. Global warming and Africa climate change. In: Low PS, editor. Climate Change and Africa. Cambridge: Cambridge UP; 2005. pp. 29-40

[17] Kurukulasuriya P et al. Will African agriculture survive climate change? World Bank Economic Review. 2006;**20**: 367-388

[18] van de Steeg A et al. The Influence of Climate Variability and Climate Change on the Agricultural Sector in East and Central Africa. Sensitizing the ASARECA Strategic Plan to Climate Change, Research report 22. Nairobi, Kenya: International Livestock Research Institute (ILRI); 2009

[19] Conway D et al. GCM simulations of the Indian Ocean dipole influence on East African rainfall: Present and future. Geophysical Research Letters. 2007; **34**(3):1-6

[20] Dore M. Climate change and changes in global precipitation patterns: What do we know? Environmental International. 2005;**31**(8):1167-1181

[21] Haile M. Sub-Saharan Africa weather patterns, food security and humanitarian response in Sub-Saharan Africa. Philosophical Transactions of Royal Society B, Biological Science. 2005;**360**: 2169-2182

[22] Hulme M et al. African climate change: 1900–2100. Climate Research. 2001;**17**(2):145-168

[23] Nicholson S, Some B, Kone B. A note on recent rainfall conditions in West Africa, including the rainy season of the 1997 El Nin˜o year. Journal of Climate change. 2000;**13**(14):2628-2640

[24] Vogel C, O'Brien K. Climate forecasts in Southern Africa. In: O'Brien K, Vogel C, editors. Coping with Climate Variability: The Use of Seasonal Climate Forecasts in Southern Africa. Surrey: Ashgate; 2003. pp. 3-35

[25] Müller C. Climate Change Impact on Sub-Saharan Africa: An Overview and Analysis of Scenarios and Models. Bonn: Deutsches Institut für Entwicklungspolitik (DIE), Discussion Paper 3; 2009

[26] McSweeney C, New M, and Lizcano G. UNDP climate change country profiles: Tanzania. 2006. Available from: https://www.geog.ox.ac.uk/research/climate/projects/undp-cp/UNDP_reports/Tanzania/Tanzania.lowres.report.pdf [Accessed: May 13, 2019]

[27] Harris R, Todaro M. Migration, unemployment and development: A two-sector analysis. American Economic Review. 1970;**60**(1):126-142

[28] Azam P, Gubert F. Migrants' remittances and the household in Africa: A review of the evidence. Journal of African Economies, AERC Supplement. 2006;**2**(15):426-462

[29] Taylor E. The new economics of labour migration and the role of remittances in the migration process. International Migration. 1999;**37**(1):63-88

[30] Gray C, Bilsborrow R. Environmental influences on migration in rural Ecuador. In: Paper for the 2010 annual meeting of the Population Association of America. Dallas; 2010

[31] Munshi K. Networks in the modern economy: Mexican migrants in the U. S. Labor market. Quarterly Journal of Economics. 2003;**118**(2):549-599

[32] Feng S et al. Linkages among climate change, crop yields and Mexico–US cross-border migration. Proceedings of the National Academy of Sciences. 2010; **107**(32):14257-14262

[33] Kniveton D, et al. Climate change and migration: Improving methodologies to improving methodologies to estimate flows. International Organization for

Migration (IOM). Geneva: Migration Research Series No.33, IOM; 2008

[34] Ezra M, Kiros G. Rural out-migration in the drought prone areas of Ethiopia: A multilevel analysis. International Migration Review. 2001; **35**(3):749-771

[35] Meze-Hausken E. Migration caused by climate change: How vulnerable are people in dryland areas? Mitigation and Adaptation Strategies for Global Change. 2000;**5**(4):379-406

[36] Naudé W. Conflict, disasters and no jobs: Reasons for international migration from Sub-Saharan Africa. Helsinki: UNU-WIDER, Research Paper No. 85; 2008

[37] Afifi T, and Warner K. The impact of environmental degradation on migration flows across countries. Working Paper No.5/2008, Bonn: UNU-EHS; 2008

[38] Reuveny R, Moore H. Does environmental degradation influence migration? Emigration to developed countries in the late 1980s and 1990s. Social Science Q. 2009;**90**(3):461-479

[39] Drabo A, Mbaye L. Climate Change, Natural Disasters and Migration: An Empirical Analysis in Developing Countries. IZA Discussion Paper No. 5927. Bonn: Institute for the Study of Labor (IZA); 2011

[40] Gray L, Mueller V. Natural disasters and population mobility in Bangladesh. Proceedings of National Academic of Science. 2012;**109**(16):6000-6005

[41] Paul B. Evidence against disaster-induced migration: The 2004 tornado in north-central Bangladesh. Disasters. 2005;**29**(4):370-385

[42] Nawrotzkia R, Maryia B. Climate and international migration in Burkina Faso and Senegal: Exploring the agricultural pathway. In: Manuscript Submitted for Consideration to be Presented at the 2016. Washington, DC: Annual Meeting of the Population Association of America (PAA); 2016

[43] Schlenker W, Roberts J. Nonlinear temperature effects indicate severe damage to U.S. crop yields under climate change. Proceedings of the National Academy of Sciences. 2009;**106**(37): 15594-15598

[44] Tanzania National Bureau of Statistics (TNBS). (2014). National Panel Survey (NPS 2012-2013)

[45] Morriset J. Tanzania Economic Update: Opening the Gates. Washington: World Bank; 2013

[46] World Bank. (2012). The World Bank Living Standard Measurement Survey (LSMS). Available from: http://surveys.worldbank.org/lsms/programs/integrated-surveys-agriculture-ISA/tanzania#bootstrap-panel–4 [Accessed: May 13, 2019]

[47] Stark O, Levhari D. Migration and risk in less developed countries. Economic Development and Cultural Change. 1982;**31**(1):191-196

[48] Stark O, Bloom D. The new economics of labor migration. American Economic Review. 1985;**75**(2):173-178

[49] Lucas E, Stark O. Motivations to remit: Evidence from Botswana. Journal of Political Economy. 1985;**93**(5):901-918

[50] Katz E, Stark O. Labor migration and risk aversion in less developed countries. Journal of Labor Economics. 1986;**4**(1): 134-149

[51] Climate Research Unit (CRU). The University of East Anglia, Gridded

climate data. 2015. Available from: http://www.cru.uea.ac.uk/data [Accessed: May 13, 2019]

[52] de Janvry A et al. Delinking land rights from land use: Certification and migration in Mexico. American Economic Review. 2015;**105**(10): 3125-3149

[53] de Brau A, Mueller V. Do limitations in land rights transferability influence mobility rates in Ethiopia? Journal of African Economies. 2012;**21**(4):548-579

[54] Feng S, Heerink N. Are farm households' land renting and migration decisions inter-related in rural China? NJAS—Wageningen Journal of Life Sciences. 2008;**55**(4):345-362

[55] ESRI. ArcGIS 10.1 Projection Engine Tables: Geographic and Vertical Coordinate Systems, Datums and Vertical Datums, Spheroids, Prime Meridians, and Angular Units of Measure. Environmental Systems Research Institute. California: Redlands; 2012

[56] Schwartz A. Interpreting the effect of distance on migration. Journal of Political Economy. 1973;**81**:1153-1169

[57] Lucas R. International Migration Regimes and Economic Development. 2005. Available from: http://www.egdi.gov.se/migration.htm [Accessed: October 10, 2017]

[58] Mendola M. Migration and technological changes in rural households: Complements or substitutes? Journal of Development Economics. 2005;**85**(1–2):150-175

[59] Borooah VK. Logit and Probit: Ordered and Multinomial Models. Quantitative Applications in the Social Sciences. Thousand Oaks: Sage; 2002. pp. 07-138

[60] Wildan S. Determinants of Labor Migration Decisions: The Case of East Java, Indonesia, International Labour Migration Series No.12. Germany: University of Kassel; 2012

[61] Bezabih M, et al. Participation in Off-Farm Employment, Rainfall Patterns, and Rate of Time Preferences: The Case of Ethiopia. vol. 10(21) Environment for Development (EFD), Discussion Paper Series; 2010

[62] Holden ST, Deininger K, Ghebru H. Tenure insecurity, gender, low-cost land certification and land rental market participation. Journal of Development Studies. 2011;**47**(1):31-47

Section 3

Land Use and Land Cover Changes

Chapter 4

Dynamics, Anomalies and Boundaries of the Forest-Savanna Transition: A Novel Remote Sensing-Based Multi-Angles Methodology Using Google Earth Engine

Alfred Homère Ngandam Mfondoum,
Igor Casimir Njombissie Petcheu,
Frederic Chamberlain Lounang Tchatchouang,
Luc Moutila Beni, Mesmin Tchindjang
and Jean Valery Mefire Mfondoum

Abstract

This chapter proposes a remote sensing multi-angles methodology to assess the transition at the interface of the forest-savanna land cover. On Sentinel2-A median images of successive dry seasons, three referential and nine analytical spectral indices were computed. The change vector analysis (CVA) was performed, selecting further one magnitude per index. The averaged moving standard deviation index (aMSDI) was proposed to compare spatial intensity of anomalies among selected CVA, and then statistically assessed through spatial and no-spatial autoregression tests. The cross-correlation and simple linear combination (SCL) computations spotted the overall anomaly extent. Three machine learning algorithms, i.e., classification and regression trees (CART), random forest (RF), and support vector machine (SVM), helped mapping the distribution of each specie. As result, the CVA confirmed each index ability to add new information. The aMSDI gave the harmonized interval [0–0.083] among CVA, confirmed with all $p - values = 0, z - scores > 2.5$, clustering of anomaly pixel,and adjusted $R^2 \leq 0.19$. Three trends of vegetation distribution were distinguished with 88.7% overall accuracy and 0.86 kappa coefficient. Finally, extremely affected areas were spotted in upper latitudes towards Sahel and desert.

Keywords: Forest-savanna, Google earth engine, Sentinel2-a, change vector analysis, spatial dynamics, averaged moving standard deviation index, autoregression tests, machine learning

1. Introduction

The global structure and productivity of ecosystems are deeply impacted by joined climate conditions and human drivers, causing general vegetation degradation [1]. The phenomenon has kept increasing in the last four decades and eventually affects whole ecosystem, soil productivity, biological systems, biotic diversity, and other environmental systems' ability to support human needs in concerned areas. Main indicators are the decline in parameters, such as low biomass, less ecological production, fragmentation, or lower canopy cover [2, 3].

Inside the tropics, vegetation is globally sensitive to seasonal and inter-annual variation in precipitations and temperatures. Extremes seasonally, i.e., longer rainy season and shorter dry season in lowest latitudes, versus the reverse phenomenon towards medium latitudes, influence the vegetation distribution with several phenological and physiological adaptations, including cover and status changes [4–6]. Typically, forest colonizes wetter areas, while savannas cover drier areas, with a gradual species distribution such as dense forest, tree savannas, grassy/herbaceous savannas, and isolated desert shrubs or clumps of dry grasses known as steppes. However, transitions are not rigidly determined by climate [7]. There is an extensive overlap between forest and savanna creating a mosaic of landscapes, and most studies on the subject remain widely hypothesized and modeled with controversial results, supported by questionable evidences. Biases include the high species turnover around 1000 mm to 2500 mm rainfall, the (un)stable states of forest and savanna maintained by feedbacks between tree cover and disturbances, and for the satellite-based approaches, the structural (in)difference between trees or grasses layer [8].

These specificities are challenging to spatialize at a point that sub-Saharan African ecosystems have played a key role in the development of remote sensing of vegetation for decades [9–11]. Nowadays, several satellite-based models provide scalable spectral information relevant to vegetation distribution and changes, physiology, and phenology, in broad terms, to monitor and combat land degradation, especially in African drylands [12–15]. As such, numerous spectral indices measure the vegetation parameters [16]. The Normalized Difference Vegetation Index, NDVI, especially, has purposely been widely used [17]. However, some limitations like sensitivity to soil background effects and atmospheric influence as well as values saturation under dense and multilayered canopy, usually alter the NDVI capacity to simultaneously predict senesced vegetation and efficiently discriminate individual anomalies, i.e., growth, vigor, leaf area index, biochemical components (anthocyanin, carotenoids, cellulose, etc.), water content or pigmentation [18–20] with accuracy. Then some previous studies focused on identifying or modeling the direction of change as well as underlying drivers of drylands vegetation [21]. Those models applied to two or more spatially close and interwoven vegetation species, require to implementation of specific processings. To the best of our knowledge, the recent progress in modeling sub-Saharan vegetation transition introduced the term of "bistability" around lower and upper transition boundaries between forest and savanna [8]. This model is based on paleo-ecological evidences (soil, topography) and climatic parameters change and oscillation (rainfall and temperature), that influence (for the firsts) and predispose (for the seconds) these two species to coexistence. With the support of a floristic survey, the ambiguity of mischaracterizing savanna as a degraded forest was clarified at some point, by identifying, the dense forest, the "bistable" forest, the "bistable" savanna, and the proper savanna.

This study approaches the forest-savanna transition study, by investigating its different spectral behaviors and their statistical meaning, in a context lacking field data, and from an open-source/open-data perspective. The triple aim is to assess the dynamics, discriminate species disturbance based on their empirical spatial distribution, and predict their extent and boundaries. As such, assuming a blurred boundary and an overlapping spatial gradient between the two species, some phenological and physiological characteristics are considered as separately as possible in terms of anomalies, and further integrated beneath the same model, so as to locate the spots requiring permanent monitoring or sustainable actions, without mischaracterizing punctual changes, factual distribution, and most accurate delineation.

2. Materials and methods

2.1 Study area

The study was conducted on the sub-Saharan mixed ecoregion, highly dependent on varied annual precipitations (AP) and yearly medium to high average temperatures (T^0), which both influence relative humidity (RH) changes. The area belongs to the medium Cameroon (central Africa), between latitudes $5^0$0'-$8^0$5'N and longitudes $10^0$0′-$15^0$5′E, a climatic transition between the agroecological zones of western highlands (AP = 1800-2500 mm; T^0 = 19.5°C; RH = 75%), bi-modal rainforest (AP = 1700-2000 mm; T^0 = 24°C; RH = 80%), and then Guinean high savanna (AP = 1500–1800 mm; T^0 = 30°C; RH = 60%) to Sudano-Sahelian savanna (AP = 400–1200 mm; T^0 = 28°C; RH = 50%) for the core area. Specifically, the vegetation density broadly reflects the climate gradient of dense moist broadleaf forest and highland forest to sparse extensive savannas featuring the co-dominance of woody shrubs, grassland in plains, and herbaceous steppe at the edge of Sahel (**Figure 1**) [22].

2.2 Data and working environment description

The study was conducted using European Spatial Agency, ESA, Sentinel2-A multispectral instrument (MSI) data, which represents a very valuable opportunity for the fine characterization and monitoring of vegetation types on large scales, but is poorly investigated for the tropical biome study [11, 23]. This sensor provides 13 varied spectral bands from 0.443 to 2.190 micrometers, a 10-day repeat cycle, and a spatial resolution up to 10 meters (Additional material 1).

The phenological dry season, globally from November to March, was selected because of its high temperatures and less rainfalls, assuming they are ideal conditions to observe the vegetation adaptations to extremes, for years 2015 to 2021. In the GEE cloud coding environment and using the JavaScript opensource simplified coding, the median reducer function was appended as the pixel-wise computation of all bi-annual collection images, based on a band per band processing [24]. Then, applying a date filter from November 15 to end of March 31, a boundary filter for the study area, and a cloud cover acceptance filter below 10%, one image of 13 bands was outputted per bi-annual periods 2015–2016 to 2020–2021, displayed a,nd converted from 16 bits to 8 bits before further processings. Offline tools, i.e., desktop software, were used to extract statistics and for some complementary processings using less memory, including final layouts. Namely, Erdas Imagine 2020, ArcGIS Pro 2020, and Microsoft Excel with extension Xlstat were specifically used.

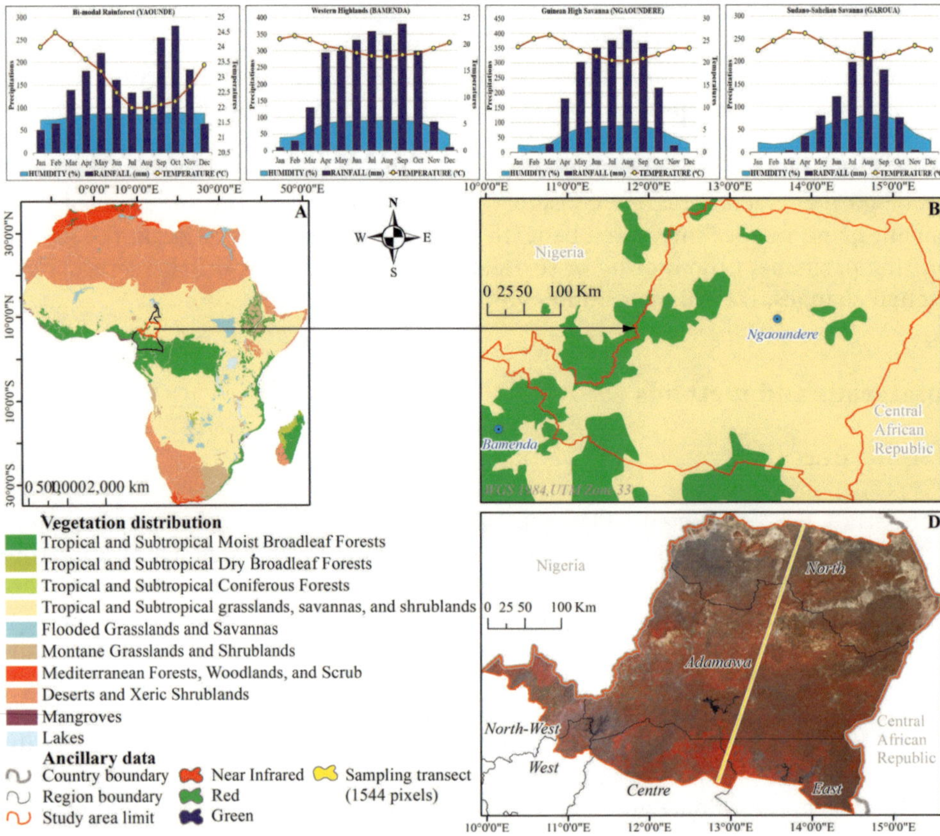

Figure 1.
Location of the study: (A) Cameroon in the context of African ecoregions based on Olson et al. (2001); (B) distribution of ecoregions in Cameroon; (C) preview of the subset ecoregions; (D) Sentinel2-a median image of the subset for mid-November 2020 to end march 2021.

2.3 Inter-seasonal dynamics, spectral assumptions, and casting of indices

In arid and semi-arid regions, common changes in density, spatial distribution, chlorophyll, pigmentation, water stress, anthocyanin, nitrogen, carotenoids, leaf structure, and browning or senescence differently impact the biomass [25]. Previous spectral indices-based applications investigated that, the visible (0.4–0.7 μm) wavelengths respond to photosynthetic and non-photosynthetic pigments, the NIR (0.7–1.4 μm) wavelengths respond to the cellular structure and exhibit solar-induced fluorescence (SIF) and the SWIR (1.4–3 μm) wavelengths respond to senescent non-photosynthetic vegetation. As such, the anisotropic behavior of vegetation at visible-SWIR wavelengths has been parameterized to describe vegetation structure [26].

We computed twelve spectral indices, whose three were selected as reference data according to their ability to better highlight the land cover targeted, i.e., second modified soil adjusted vegetation index, MSAVI2 [27], to assess the vegetation cover, normalized difference soil drought index NDSoDI [14], to highlight the dry bare soils, and new water index, NWI [28], to map the surface water. After testing dozens of other indices, two assumptions were emitted for an efficient last casting, such as *i) two or more indices targeting the same anomaly, with different or improved (increase/decrease) spatial patterns, express new information to be consider*; *ii) two or more indices targeting a different anomaly should show different spatial patterns to be considered enough informative*. Accordingly, nine vegetation indices were selected and computed, i.e., photosynthetic vigor index (PVR), global vegetation moisture index (GVMI), plant biochemical index (PBI), first red-edge inflection point (REIP1), modified chlorophyll reflectance index green (mCRIG), leaf water content index (LWCI), modified anthocyanin reflectance index (MARI), simple ratio pigment index (SRPI), and plant senescence reflectance index (PSRI). **Table 1** gives details of their formulations with references.

These indices were stacked and previewed with, MSAVI2, NDSoDI and NWI seeking the following: *i) agreement or disagreement with MSAVI2*; *ii) absolute disagreement with NDSoDI*; *iii) disagreement when possible NWI* (Additional material 2). On the transversal transect of 1544 pixels covering all types of vegetation and land cover, 500 pixels were sampled on different sections (100/300 pixels) for each index, to compare trends and relationships with MSAVI2 and NDSoDI. The two periods moving average was used to better visualize trends, while the simple linear regression was performed to extract the coefficient of correlation, R^2, and the root mean square error, RMSE (**Figure 2**). Based on patterns distribution and these statistics, a threshold was defined for each index, to separate vegetated and non-vegetated areas (Additional material 3). The binarized images were used as entries for the change assessment.

Name	Equation	Primary goal	Reference
MSAVI2	$\frac{2NIR+1-\sqrt{(2NIR+1)^2-8(NIR-Red)}}{2}$	Map vegetation cover while reducing soil background	[27]
NDSoDI	$\frac{(Red+SWIR1)-Blue}{(Red+SWIR1)+(Blue+L)}(1+L)$, $with\ L : [0.4]$	Map dry soils while reducing surroundings source of moisture	[14]
NWI	$\frac{Blue-(NIR+SWIR1+SWIR2)}{Blue+(NIR+SWIR1+SWIR2)}$	Extract surface water	[28]
PVR	$\frac{Green-Red}{Green+Red}$	Photosynthetic vigor for crop monitoring	[29]
GVMI	$\frac{(NIR+0.1)-(SWIR1+0.02)}{(NIR+0.1)+(SWIR1+0.02)}$	Vegetation water content and evapotranspiration	[30]
PBI	$\frac{NIR}{Green}$	General biochemical reflectance	[31]
REIP1*	$700+40\left(\frac{\left(\frac{Red+Rededge3}{2}\right)-Rededge1}{Rededge2-Rededge1}\right)$	Canopy chlorophyll, nitrogen content and polluted soil dynamics	[32]
mCRIG	$(Blue^{-1}-Green^{-1})NIR$	Chlorophyll, carotenoids and anthocyanin	[33]
LWCI	$LWCI=\frac{\log 1-(NIR-SWIR1)}{-\log 1-NIR-SWIR1}$	Water content, change and stress	[34]
MARI	$(Green^{-1}-Red^{-1})NIR$	Chlorophyll, carotenoids and anthocyanin	[35]
SRPI**	$\frac{DeepBlue}{Rededge2}$	Nitrogen content, water and chlorophyll	[36]
PSRI	$\frac{Red-Blue}{NIR}$	Pigmentation and vigor changes dues to vegetation senescence	[37]

**Was inverted after first preview, to better highlight vegetation patterns instead of soil dynamics.*
***Was adapted to SENTINEL2-A covered wavelengths by using band 6.*

Table 1.
Spectral indices used.

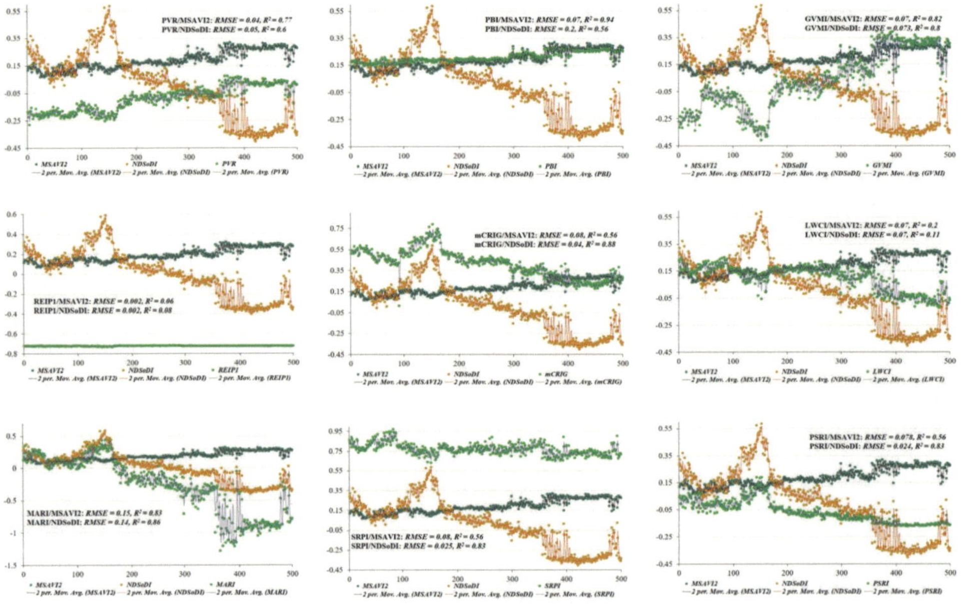

Figure 2.
Relationships between each analytical spectral index and both referential ones. The bright green curve represents the dependent variable, while the two others are explanatory variables (MSAVI2 = dark green; NDSoDI = light brown). N = 500 for all variables and regression parameters. The following is noticed from the trends of curves: PVR, PBI and GVMI have a neat positive correlation with MSAVI2, but a sharp negative one with NDSoDI; conversely, MARI and PSRI curves describe the exact opposite trends (positive with NDSoDI and negative with MSAVI2); mCRIG and LWCI curves evolve in another direction cutting the MSAVI2 and NDSoDI curves, while IREIP1 curve shows no real relationship them.

2.4 Change vector analysis and spatial dynamics assessment

This process creates a difference image between two or more bands in a multi-temporal image analysis, so to detect changes in the type or the conditions of surface features. Depending on the study, change vector analysis (CVA) can use calculation principles of the Mahalanobis or the Euclidean distance as in this study, according to the following expression [38]:

$$\boldsymbol{R} = \sqrt{(\beta\sigma_1 - \beta\sigma_2)^2 + (\beta\rho_1 - \beta\rho_2)^2} \quad (1)$$

Where $\boldsymbol{R}$ is magnitude of the vector change, $\boldsymbol{\beta\sigma_1}$ and $\boldsymbol{\beta\sigma_2}$ are fraction in date 1 and date 2 respectively, $\boldsymbol{\beta\rho_1}$ and $\boldsymbol{\beta\sigma_2}$ are fraction cover in date 1 date 2.

Four classes of magnitude are represented for either degradation or re-growing [39] (Additional material 4). For each binarized image, a CVA process was performed, the magnitudes of increase and decrease patterns were used, while the stability magnitudes were ignored. A total of twelve difference images per index resulted, i.e., six per selected magnitude. Considering the need of complementarity among indices, only one CVA magnitude, i.e., increase or decrease, was selected per index for further processing. Criteria used to optimize the selection were the original goal of each index, as well as its spatial and statistical relationship with MSAVI2, NDSoDI and NWI (**Table 2**).

Index	Targeted asset	Compared trends to MSAVI2 / NDSoDI / NWI	CVA
PVR	Age & Cover	Same with a plus / Inverted / Inverted	Decrease
GVMI	Water content & Cover	Same with a plus / Inverted / Inverted	Decrease
PBI	Inner composition & Cover	Same with a plus / Inverted / Inverted	Decrease
IREIP1	Height, Cover & Health	Different / Inverted / Inverted	Decrease
mCRIG	Health & Composition	Different / Inverted / Inverted	Increase
LWCI	Water stress	Different / Inverted / Close	Decrease
MARI	Health & Inner composition	Inverted / Close / Inverted	Increase
SRPI	Health & Inner composition	Inverted / Close / Inverted	Increase
PSRI	Age & Cover	Inverted / Close / Inverted	Increase

Table 2.
CVA magnitude selected per spectral index.

2.5 Averaged moving standard deviation index method and anomalies mapping

The moving standard deviation index, MSDI, is a filter applied to satellite images multispectral or derivative channels using the moving standard deviation calculation, generally to assess degradation [40]. One common application is the vegetation and soil of semi-arid systems, where the variability of the MSDI is used to indicate levels of habitat degradation [41–43]. MSTDI has been proven efficient to operate well in complex regions [40, 42, 44].

Here, the five derivative images per selected CVA were used as entries. A standard deviation was computed for each entry. Difference between consecutive standard deviations were calculated, giving four new images. Then, the averaged MSDI (aMSDI) was conceived as follows:

$$aMSDI = \frac{\sum_{i=1}^{n} \sigma_i b_i}{n + \tau} \tag{2}$$

For all $1 \quad i \quad n$, $\sigma_i b_i$ represent the ith CVA-band contribution to the information distribution in the outputted image, whereas τ is the difference between the number of times each index was originally calculated (7 times, starting from the multispectral median images) and the number of resulting standard deviations computed among selected CVA-bands (5). Here, $\tau = 2$. Computations were performed at 3×3 -, 5×5- and 7×7-pixel moving window. The same steps were applied to both of decrease and increase CVA magnitudes of MSAVI2. The goal at this point was to assess spatial convergence and divergence trends with the nine analytical vegetation spectral indices' aMSDI, as a first-hand validation process of anomalies distribution. Two autoregression-based tools were used as metrics at this step (ESRI, 2014):

- The global Moran's I index was computed to assess the spatial autocorrelation of aMSDI outputs along the 1544 pixels transect. Its integration in spatial analysis is important to avoid incorrect statistical inference from inefficient or biased parameter estimates [45]. The formula is expressed such as follows:

$$I = \frac{n}{S_O} \frac{\sum_{i=1}^{n} \sum_{j=1}^{n} w_{i,j} z_i z_j}{\sum_{i=1}^{n} z_i^2} [-1, 1] \tag{3}$$

Where z_i is the deviation of an attribute for pixel i from its mean $x_i - \overline{X}$, $w_{i,j}$ is the spatial weight between pixel i and j, n is equal to the total number of pixels, and S_O is the aggregate of all the spatial weights, developed as:

$$S_O = \sum_{i=1}^{n} \sum_{j=1}^{n} w_{i,j} \tag{4}$$

Values closer to **0** indicate little or no spatial autocorrelation, regular values indicate a clustered trend, and negative values indicate a dispersed trend. In this paper and based on the proposed model, i.e., one CVA magnitude per index and average of the MSDI among years, the clustered trends were targeted to express values closeness. The null hypothesis for the patterns analysis was emitted for a less randomness and a significant clustering, comparing how good would be the model to spot and regroup spatially related pixels of anomalies, and discriminate them from "healthier" pixels. High standard deviation (all $z - scores > 2.5$) and low probability (all $p - values = 0$) were ranges targeted, meaning that the observed spatial patterns are probably too unusual to be the result of random chance.

- Ordinary Least Squares (OLS) regression was computed as the non-spatial measure of the spatial convergence/divergence of aMSDI. This method predicts or models a dependent variable relationship with a set of explanatory variables. The regression coefficients are usually estimated by using least-square techniques such as:

$$y = \alpha + \sum_{j=1}^{n} x\beta_j + \varepsilon \tag{5}$$

Where, α is the intercept, β_j is the regression coefficient, and ε is the residual term, given by the difference between observed and expected value. Under these assumptions, the regression coefficient can be obtained by [46]:

$$\hat{\beta} = \left(X^T X\right)^{-1} X^T y \tag{6}$$

Here, each aMSDI of the nine indices was considered as dependent variable, whereas MSAVI2's aMSDI of both CVA, were simultaneous explanators. Three OLS parameters were selected for interpretation, i.e., adjusted R-squared, corrected Akaike's information criterion (AICc) which assesses the best-fit model between spatial and non-spatial Ordinary Least Squares (OLS) models, root mean squares error (RMSE), and three graphs, i.e., scatterplots of the relationship among variables, histograms of standard deviation probability and plots of residuals versus predicted.

At this point, the spatial correlation was assessed through the cross-correlation mapping process using only the finest 3×3 -pixel moving window, that consistently shows best statistic performances than the two others [43]. The parameters of the algorithm were set for the maximum gap (maximum pixel shifting) at **1**, and the maximum masked fraction (maximum fraction of the pixel within the correlation window that are allowed to be masked) at $\mathbf{0}.5$.

Further, the individual maps of anomalies were used to produce a single one. The first step consisted in stacking the aMSDI for each $\mathbf{3} \times \mathbf{3}$ -pixel moving window. Then, another correlation map was computed for all the nine inputs, only selecting the correlation coefficient layer. The highest values of the outputs were assumed show agreement and disagreement among aMSDI for high and low anomalies. At last, to keep the highest values as signs for anomaly spots and the lowest ones as potentially "healthier" areas, a simple linear combination, SLC, was performed among the nine aMSDI for each moving window size, by summing the layers without a weighing factor.

2.6 Machine learning for vegetation discrimination

Based on visual trends, binarized images were regrouped in three axes of three indices of vegetation each. A principal component analysis, PCA, was performed for each ax using the covariance reducer algorithm, that reduces some number of $\mathbf{1} - \boldsymbol{D}$ arrays of the same length $\boldsymbol{N}$ to a covariance matrix of shape $\boldsymbol{N} \times \boldsymbol{N}$. This reducer uses the one-pass covariance formula [24]. The first component of each output was chosen, and a threshold was set to separate vegetated and non-vegetated areas. Then the three outputs were stacked with NDSoDI and NWI to form a new five bands image. In Ref. to both this new image and to the last study period median image (2021–2022), three classes of vegetation were defined, completed by soil/Built-up and water classes. At least two spectral curves were produced per class to confirm the trends, except for water that appeared uniform. The land use land cover, LULC, classes sampling was performed using 115 points, and a process of $[\mathbf{1} : \mathbf{1}]$ ratio between training samples on the PCA stacked image and testing samples on the MS image (**Table 3**).

Three machine learning algorithms were performed, such as the Classification and Regression Trees (CART), Random Forest (RF) and Support vector machine, (SVM) [47–49]. The metrics used to assess general performance of each classifier were overall accuracy, OA, and the kappa coefficient, KC [50, 51]. Whereas, the thorough assessment of their efficiency on individual LULC classes was done using the error matrix to extract the producer accuracy (PA) and the user accuracy (UA). Eqs. (7) and (8) formulate the main quality measures of the learning:

$$\mathrm{OA} = \frac{\textit{Total number of correct samples}}{\textit{Total number of samples}\ (\%)} \tag{7}$$

$$\mathrm{KC} = \frac{\varepsilon_1 - \varepsilon_2}{1 - \varepsilon_2} \tag{8}$$

Training:Testing					Total
Vegetation1	Vegetation2	Vegetation3	Soil/Built-up	Water	
100:100	100:100	100:100	100:100	100:100	500:500

Table 3.
Samples distribution.

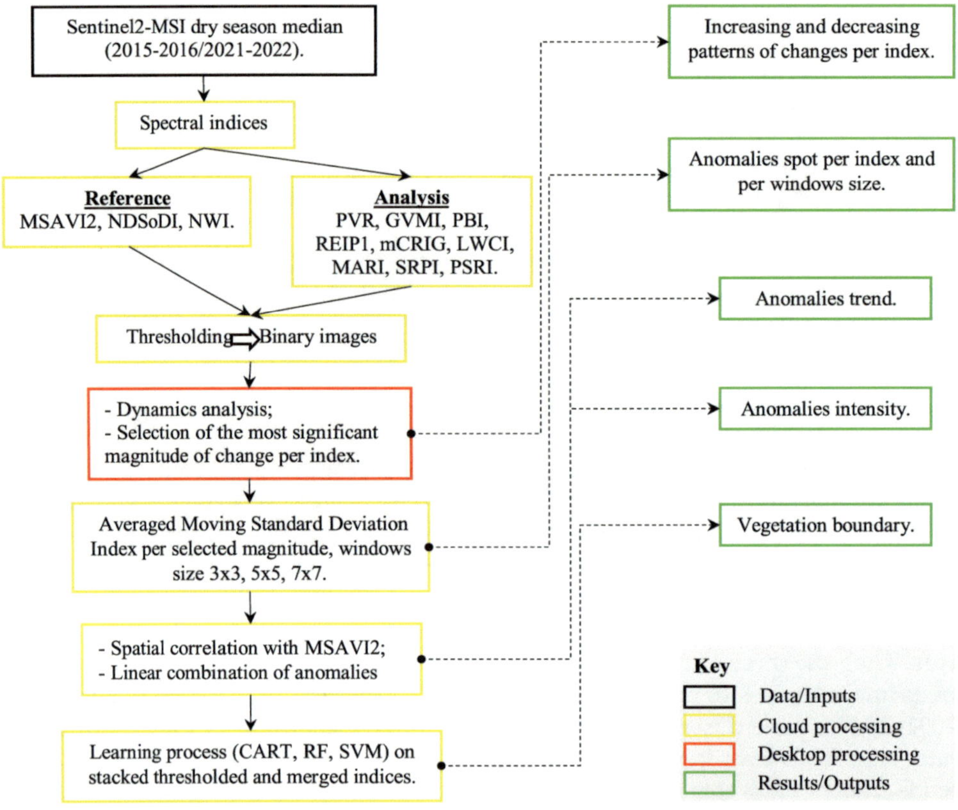

Figure 3.
Overview of the study design.

with,

$$\varepsilon_1 = \frac{\sum_{i=1}^{n} D_{ii}}{N}, \text{ and, } \varepsilon_2 = \frac{\sum_{i=1}^{n} D_{i+} D_{+i}}{N^2}$$

Where, D_{ii} is the number of observations in row i and column i of the confusion matrix, n is the number of rows in the error matrix, N is total number of counts in the confusion matrix, x_{i+} is the marginal total of row i, and x_{+i} is the marginal total of column i. Main steps of the processing are summarized in **Figure 3**.

3. Results

3.1 Magnitudes of spatial dynamics

Globally, each index showed a different interseason sensitivity to dynamics for the two main magnitudes of change in the vegetation distribution. The indices detecting vigor, biochemical composition, and water content, i.e., PVR, PBI and GVMI, that are highly correlated with MSAVI2, invaded the upper medium area for the increasing magnitude, while their increasing patterns were more concentrated on the south and lower medium area (**Figure 4a** and **b**). Besides, the indices detecting canopy chlorophyll, carotenoids, anthocyanin, nitrogen, and water stress, with a positive (mCRIG)

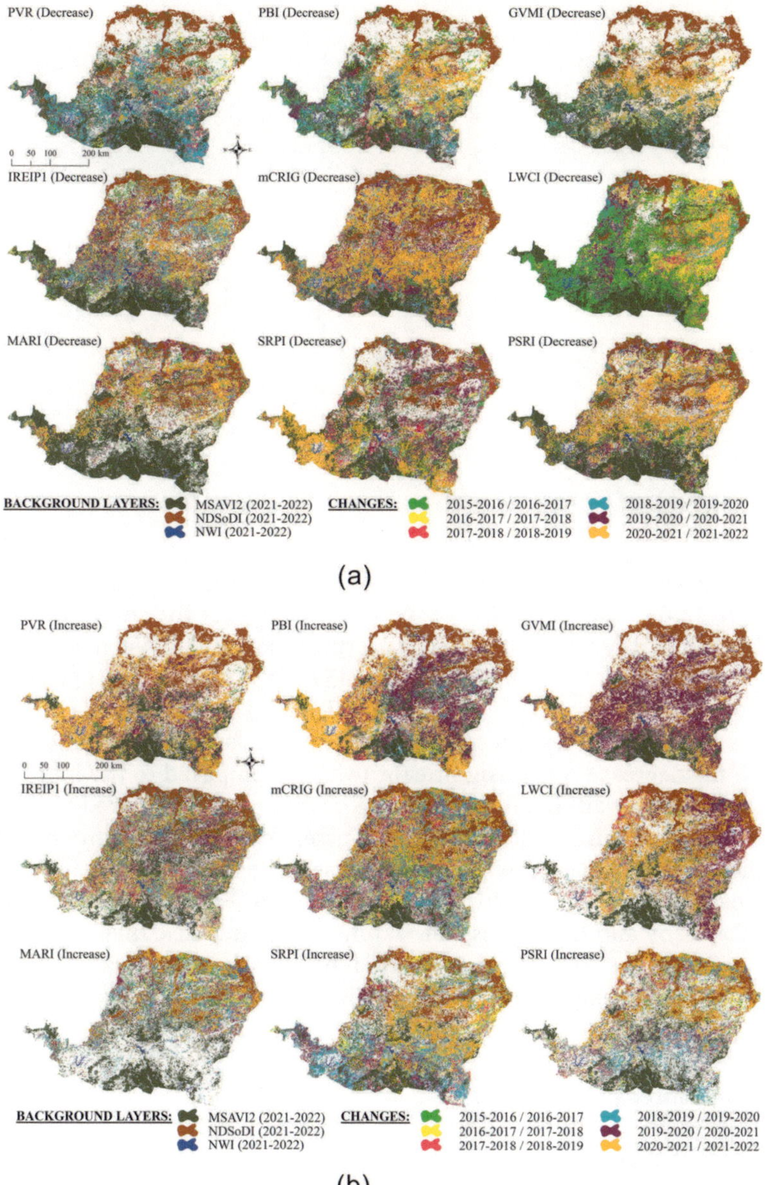

Figure 4.
a. Decrease trends of CVA per index. The background layers are from 2021 to 2022. b. Increase trends of CVA per index. The background layers are from 2021 to 2022.

or non-significant (IREIP1, LWCI) correlation to MSAVI2, gave a general decreasing or increasing patterns on the whole subset (**Figure 4a** and **b**). Whereas, the indices targeting the lack of chlorophyll, carotenoids, anthocyanin, pigment or the plant senescence, and that were negatively correlated to MSAVI2, i.e., MARI, SRPI and PSRI, cover the South area for the decreasing patterns, and the North part for the increasing patterns (**Figure 4a** and **b**).

Moreover, all magnitudes agreed to the same total areas for all indices, when adding the unchanged magnitudes. For the extreme cases, LWCI indicates a brutal

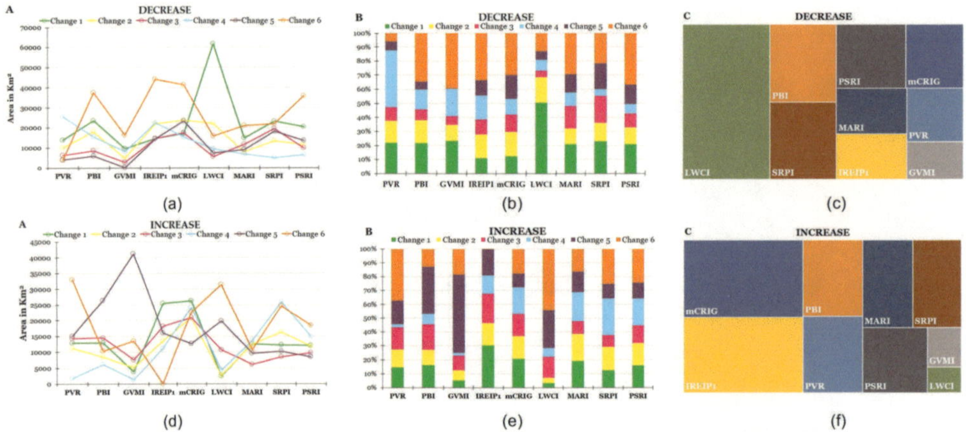

Figure 5.
Areas(a), percentages(b) and hierarchy of changes (c) based on the CVA decrease (row1) and increase (row2) magnitudes.

decrease in between 2015 and 2016 and 2016–2017 (Change1, **Figure 5a**), from 61,609 km^2 to 22,030 km^2, i.e., 58% of the total change for the study period (**Figure 5b**). While, GVMI indicates an important increase between 2019 and 2020 and 2020–2021 (Change 5, **Figure 5a**), from 1439 km^2 to 41,215 km^2, i.e., 69.4% of the total change for the study period (**Figure 5b**). Then all changes were complementary between the two magnitudes, except for IREIP1 in the last period, 2020–2021/2021–2022, which the decrease was unsignificant (4km^2) compared to the increase (**Figure 5a&b**). Above all, the magnitudes of change were compared in terms of hierarchy tree, to measure indices that better assess the spatial progression or regression. For the decrease magnitudes, LWCI assessed regression for 4/5 intervals, except during the period 2018–2019 to 2019–2020 (Change 4, **Figure 5a**), which recorded a recover followed by another regression. While for the increase magnitudes, mCRIG gave the most accurate assessment of a spatial progression for 5/5 intervals, which is the overall best score magnitudes. These statements justify their dominant position in the tree-map (**Figure 5c**).

3.2 Trends and intensity of anomalies

3.2.1 Contribution of aMSDI and spatial autoregression significance

The outputs of the aMSDI were all obtained in the same interval, **[0–0.083]** for each CVA-band, with a maximum value confirmed above $\mathbf{10^{-6}}$, in every pixel window size. It should be noticed that the binarized inputs do not interrupt the spatial gradient of the phenomena described and the values showed a continuum of intensities. Therefore, although the results are unitless, the lowest value, **[0]**, predict no potential issues, and the highest value, **[0.083]**, predict a critical intensity of the targeted degradation or anomaly. Globaly, the spatial distribution of seasonal degradation/anomaly evolves in the south-north direction (Additional material 5 a-d).

From the spatial autocorrelation model synthesized in **Table 4**, the expected Moran's I index and p-values are identical for all, −**0.000648** and **0**, respectively, while variance is almost the same for all, **0.000457** or **0.000458**. Besides, final Moran's I index varies between **0.81** for MSAVI2 aMSDI of increase CVA, and **0.95** for

aMSDI	Moran's I	Expected Moran's I	Variance	Z-Scores > +2.58	P-Values
MSAVI2_Decr	0.9	−0.000648	0.000457	42.2	0.00
MSAVI2_Incr	0.81	−0.000648	0.000457	37.9	0.00
PVR	0.95	−0.000648	0.000458	44.3	0.00
PBI	0.85	−0.000648	0.000458	39.6	0.00
GVMI	0.84	−0.000648	0.000457	39.5	0.00
IREIP1	0.76	−0.000648	0.000457	35.4	0.00
mCRIG	0.86	−0.000648	0.000457	40.3	0.00
LWCI	0.91	−0.000648	0.000457	42.5	0.00
MARI	0.91	−0.000648	0.000457	42.5	0.00
SRPI	0.86	−0.000648	0.000457	40.4	0.00
PSRI	0.84	−0.000648	0.000458	39.3	0.00

Table 4.
Spatial autocorrelation model report. N = 1544 for all variables and parameters.

PVR; whereas *z*-scores are all above **2.5**, i.e., between **37.9** for MSAVI2 aMSDI of increase CVA, and **44.3** for PVR. Therefore, the spatial patterns of every highlighted vegetation's distribution issue by each aMSDI, described an important clustering of pixels with same and closest values at a confidence level of **99%**, stating that the phenomenon is far from randomness as wished (ESRI, 2014) (Additional material 6). Added to this, the proposed model of aMSDI has been significantly stable around all parameters, while its values scale is easy to compare and interpret among outputs of different analytical spectral indices.

3.2.2 Spatial convergences of anomalies and non-spatial autoregression analysis

First outputs of the cross-correlation mapping algorithm informed on the spatial relationship between each MSAVI2's aMSDI and individual aMSDI of analytical indices. About the MSAVI2 decrease CVA, the highest positive correlation of aMSDI was with PVR (**R = 0.4; R^2 = 0.16; RMSE = 0.026**), while PBI (**R = -0.23; R^2 = 0.05; RMSE = 0.027**) and MARI (**R = -0.22; R^2 = 0.05; RMSE = 0.027**) were found totally decorrelated (**Figure 6**). As such as, the most noticeable anomaly impacting the forest-savannah vegetation cover is the decrease of vigor and other subsequent characteristics highlighted by PVR and close algorithms. Less issues might be caused by biochemical composition including anthocyanin. Concerning the MSAVI2 increase CVA, the highest positive correlation of aMSDI was established with LWCI (**R = 0.53; R^2 = 0.28; RMSE = 0.016**), at the opposite of decorrelation to MARI (**R = -0.56; R^2 = 0.32; RMSE = 0.015**) and IREIP1 (**R = -0.55; R^2 = 0.31; RMSE = 0.015**) (**Figure 6**). With its particular distribution on the East area, LWCI express that water content influence to the increase of vegetation, while anthocyanin and canopy chlorophyll have no influence.

The multi-regression performed between each analytical aMSDI and simultaneously both MSAVI2's aMSDI on **3** × **3**-pixel moving window, were consistent but specifics by case. Generally, the relationships were all highly significant (all $p = 0$). The corrected Akaike Information Coefficient (AICc) vary between -7795

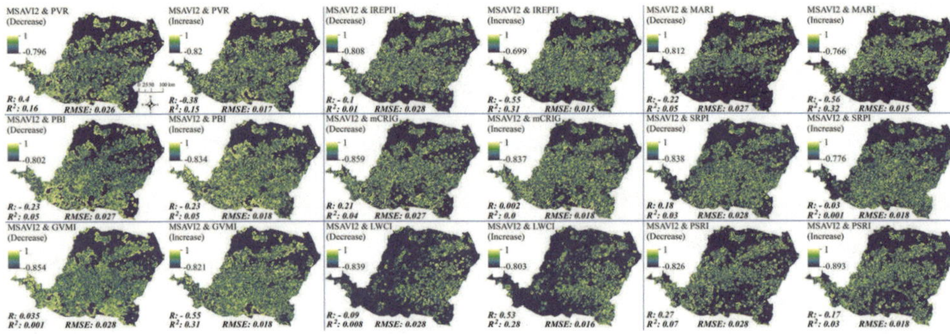

Figure 6.
Cross-correlation maps between aMSDIs of MSAVI2 and the selected aMSDI for each index.

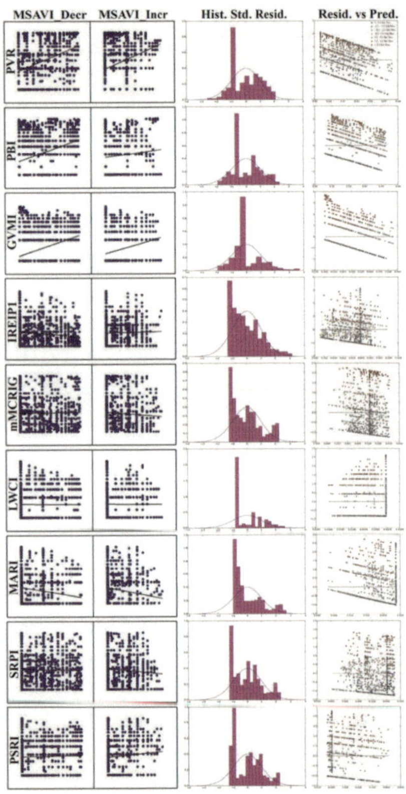

Index	Intercept	Coeff.MSAVI_Decr	Coeff.MSAVI_Incr	RMSE	Adjusted R²	ΔAICc
PVR	0.024 (0.00096)	0.369 (0.03)	0.284 (0.04)	0.028	0.17	-6642
PBI	0.02 (0.0009)	0.403 (0.028)	-0.055 (0.037)	0.034	0.13	-6853
GVMI	0.005 (0.0007)	0.33 (0.02)	0.14 (0.028)	0.027	0.19	-7706
IREIP1	0.02 (0.0007)	-0.106 (0.021)	0.077 (0.027)	0.028	0.017	-7795
mMCRIG	0.028 (0.0009)	0.046 (0.027)	0.103 (0.037)	0.03	0.005	-6903
LWCI	0.02 (0.0008)	-0.0032 (0.026)	-0.003 (0.03)	0.041	0.00002	-7121
MARI	0.024 (0.0008)	-0.14 (0.025)	-0.12 (0.034)	0.026	0.04	-7161
SRPI	0.03 (0.00085)	0.013 (0.026)	-0.103 (0.035)	0.034	0.006	-7032
PSRI	0.027 (0.0009)	-0.008 (0.027)	0.11 (0.036)	0.024	0.006	-6987

Figure 7.
OLS regression results synthesis (Eqs. (5) and (6)). All estimates have, p = 0 and n = 1544. Values in parentheses indicate standard error estimates for regression parameters. MSAVI_Decr/MSAVI_Incr = values of aMSDI for decreasing or increasing patterns of the second modified soil-adjusted vegetation index; R^2 = adjusted correlation coefficient; RMSE = root mean square error; ΔAICc = indicates the corrected Akaike information criterion, thus the spatial model improvement over the non-spatial form.

and − **6642**, for IREIP1 and PVR respectively; the intercept values are in between [**0.005–0.03**]; the regression coefficient of each MSAVI2 is more significant with indices potentially spotting high and low anomalies at the same areas, but both negatives for LWCI and MARI (**Figure 7**). The adjusted correlation coefficients were low (all R^2 **0.19**) and those of aMSDI sharing identical patterns with MSAVI2 (PVR, PBI

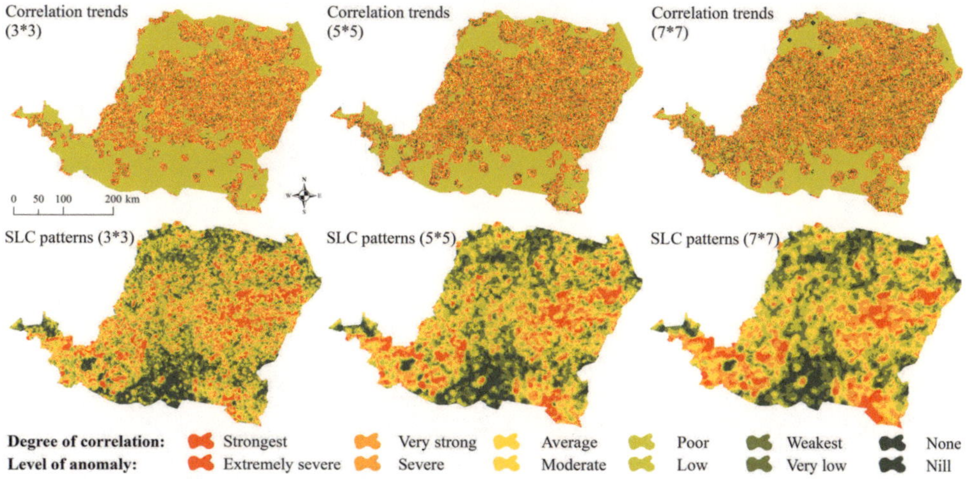

Figure 8.
Anomaly trends and spots. The synthesized correlation maps on the first row shows patchwork of high correlation in the medium area for the 3 × 3-pixel moving window, but two large continuous spots of convergence in the north-east and south-west according to 5 × 5 - and 7 × 7-pixel moving windows. The SLC outputs on the second row definitely separate high values from low values correlations.

and GVMI) recorded a normal probability distribution of standardized residuals around **0**, whereas other are skewed to the right. In both cases, there is a dominant peak, marking the separation among affected and non-affected areas, or between covered and uncovered areas. All the graphs of residuals show less randomness and more clustering per class of standard deviation (Additional material 6).

Spatial synthesis of the cross-correlation mapping algorithm confirms the distribution of patterns and global trends. The combined correlation among the nine indices' aMSDI were found strong at different degrees, for both highest and lowest values, then indicating convergences of significant or non-significant anomalies at the South-west and North-East of the area (**Figure 8**). For the three windows of calculation, they were clearly separated from predicted vegetation statuses, i.e., highly and lowly affected, soil, built-up and water features (**Figure 8**). From the cross-correlation map synthesis, the simple linear combination (SLC) outputs for each window discriminated positive from negative spatial trends. Then, the largest spots of extremely severe anomalies are located in the south-west, south-east and north-east areas, well separated on **5** × **5** and **7** × **7**-pixel moving windows. Subsequently at the finest **3** × **3**-pixel moving window, the spots of anomalies are continuously concentrated in the south-west to north-east direction (**Figure 8**).

3.3 Spatial extent and predicted delineation

The repeated sampling of vegetation gave spectral curves with at least three pairs of same trends for vegetation (**Figure 9a**). As spatial evidence, the patterns of ML were identical for the three algorithms, with three distinguished classes of vegetation, well separated from soil/built-up and water (**Figure 9b**). The three classifiers performed with a high and identical OA of **88.7%**, for a KC of **0.86** (**Figure 9b**). These measures confirm the spectral agreement between classified and multispectral image reflectance of the five land features. When confronting PA and UA, that also recorded

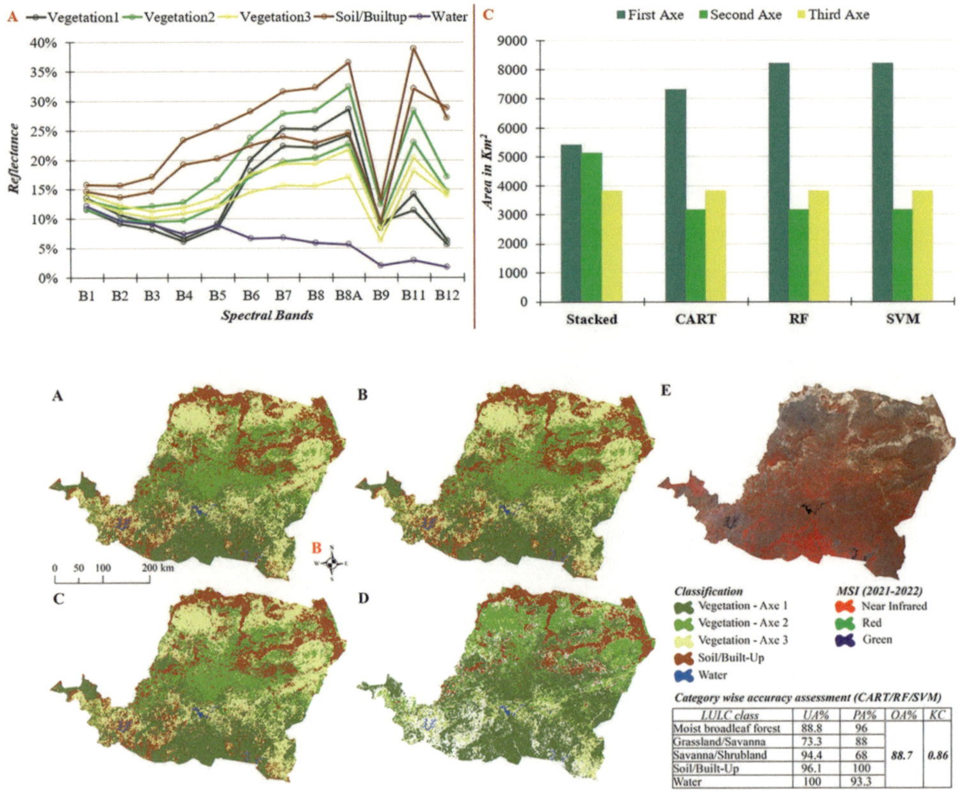

Figure 9.
Spectral curves (A), spatial patterns (B) and statistics (C) of the three axes of vegetation on LULC. In details: (A) CART; (B) RF; (C) SVM: (D) stacked axes of vegetation with soil (NDSoDI) and water (NWI) features for 2021–2022; E) Sentinel2-a image of 2021–2022.

identical values for all classifiers per class (**Figure 9b**), it can be assumed that regrouping indices thresholded images is a good option to identify the forest-savanna transition and interpenetrated vegetation. Presumably and based on the world ecoregions map as well as the empirical vegetation density, the three classes of vegetation can be identified as follow in south-north latitudes direction (Olson et al., 2001): *i) first ax matches tropical and sub-tropical moist broadleaf forests*; *ii) second ax is dominated by grasslands, mixed with savannas*; *iii) third ax mixes savannas and shrublands*.

Concerning the vegetation types, discrimination, and extent, RF and SVM gave the exact areas for the three classes, CART agreed with them for the second and third axes, whereas the stacked derivative only agreed with all the algorithms concerning the third ax, i.e., **3807.3 km^2** (**Figure 9c**). The discrepancies especially with the stacked image could be explained by the soil features footprint that is different between the NDSoDI displayed and classified image, as well as the urge overlapping noticed earlier between second and third axes. In fact, the interpenetration of the predicted grassland/savanna (axe2) and savanna/shrubland (axe3), which might better reflect the transiting vegetation behavior targeted along the study, raises on the other hand assumptions on the accurate assessment of areas. Their UA (grassland/Savanna = **73.3%** and savanna/shrubland = **94.4%**) then support this assumption (**Figure 9b**).

4. Discussions

The efficiency of the whole proposed methodology was assessed and discussed on selected aspects and the comparison with existing methods was basically empirical. At first, depending on their goals, previous MSDI-based studies analyzed only the standard deviation of the red and near infrared wavelengths, while those integrating vegetation indices were limited to three of them [40, 43, 52]. Because the goal of proposing aMSDI in this study was to assess consecutives dry season anomalies and discriminate them from empirical statuses of the forest-savanna specificities, we integrated nine spectral indices, selected on the basis of targeted phenological or physiological weaknesses, and whose computations basically integrate several wavelengths. Interestingly, although only one CVA magnitude was chosen per index, all individual models showed the expected visual convergence of similarity or dissimilarity trends.

Moreover, previous applications stated that the common calculation of MSDI on raw spectral index, gives outputs with a minimum value of zero and a maximum value determined by those of the pixels evaluated [43]. Consequently, outputs value cannot be directly compared. Here, by applying the averaging process to binarized CVA, **[0,1]**, this study alternatively addressed these issues for a multidate analysis. With same or divergent visual patterns, all outputs were scaled inside the identical interval, **[0–0.083]**, although showing convergent or divergent spatial patterns. The significance of the spatial and non-spatial autoregression models, has helped to confirm the inner variability of each aMSDI although identical scales of values and apparent same trends between others, as well as convergences/divergences of trends with the reference aMSDI (MSAVI2). Consequently, the output spotting anomalies was proof of complementary among individual contributions and spatial agreements.

Besides, common attempts of mapping distribution, typology, and delineation of forest and savanna, have always been supported by fieldwork, based on climate parameters, as well as including paleo-ecological evidences and detailed floristic survey to be efficient [8]. The methodology presented in this paper has predicted three different axes of vegetation, resulting from the PCA processing and thresholding (**Table 5**). For each of the six study periods, the first ax in the south part is composed by a dense and potentially healthier vegetation, highly correlated with the referential data MSAVI2. Whereas the other two axes, more and more sparse towards north, are divergent with the first one and somehow each with another (**Figure 10a&b**). Nevertheless, their individual trends foresee some overlapping, in the center and in the northern areas. A sampling of each ax along the transect of 1544 pixels, showed how interweaved and complexes are the boundaries among forest-savanna species (**Figure 10b**).

To answer the interrogations behind these ambiguities, a simple multicollinearity test was run, showing how independent one ax is from another. When the correlation between two independent variables is considerably high, it is a problem in the

Vegetation trends	Regrouping	PCA threshold
1st axe	PVR, PBI, GVMI	≥ 0.65
2nd axe	IREIP1, MARI, PRSI	≥ 0.95
3rd axe	mCRIG, LWCI, SRPI	≥ 0.75

Table 5.
Vegetation axes, proposed groups and PCA thresholds used to binarize.

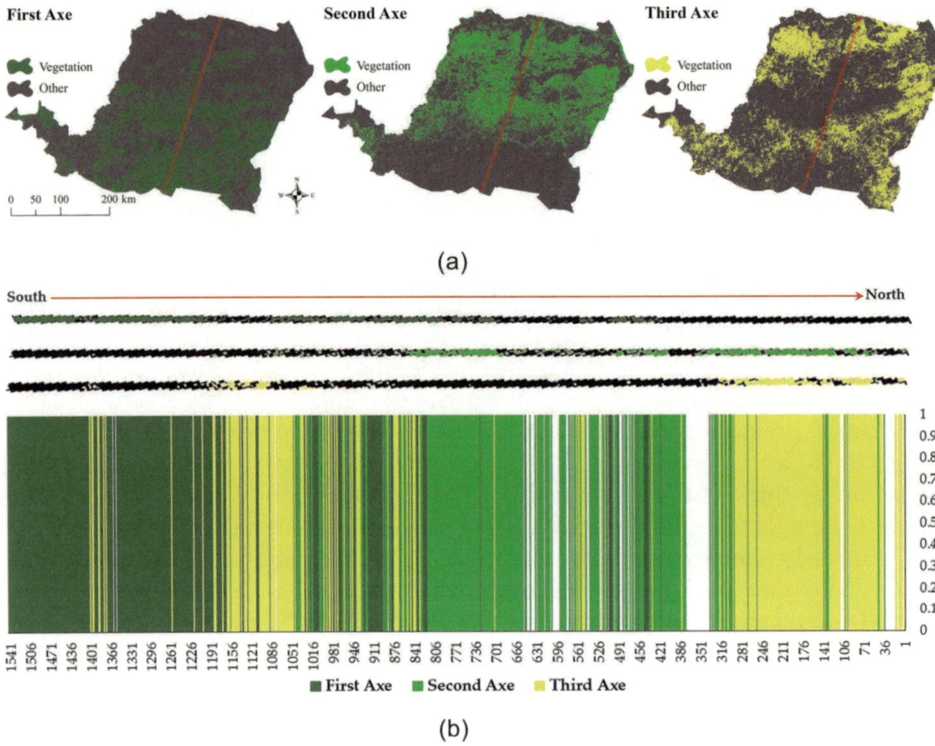

Figure 10.
a. the three main axes of vegetation's spatial distribution. b. Pixels' value along the transect of the vegetation's axes.

modeling process. The VIF (variance inflation factor) and tolerance were used for diagnosis. VIF is the reciprocal of tolerance, knowing that, tolerance is $\mathbf{1-R^2}$. We used the lowest known VIF, $\mathbf{<3}$, while expecting the highest tolerance, so to measure independence. Therefore, while the occurrence percentiles of value **1** on each ax highlights interweaving in between **[62.5–78.3]** for axes 1 and 2, and **[69.5–99.97]** for all, low VIF ($\mathbf{1.33 \le \mathit{VIF} \le 1.35}$) and high tolerance ($\mathbf{0.74 \le \mathit{Tol} \le 0.89}$) confirm the total separability, i.e., less collinearity among axes (**Table 6**).

Although from this study, we cannot properly use the qualifier of "bistable" forest or savanna, because it highly depends upon climate and paleo-ecological parameters, it is important to notice how ambiguous is the distribution and blurry are the boundaries. Thoroughly, on any ML output, three zooms distributed on three different latitudes helped to notice different types of transitions (**Figure 11**). Between the lower latitudes $5^0 30$'-$6^0 30$'North, the transition is from the first (moist broadleaf forest) to

Axe	VIF	Tolerance	Percentile for value 1
AXE1	1.33	0.75	[49.9–99.97]
AXE2	1.12	0.89	[62.5–99.97]
AXE3	1.35	0.74	[78.3–99.97]

Table 6.
Multicollinearity test results.

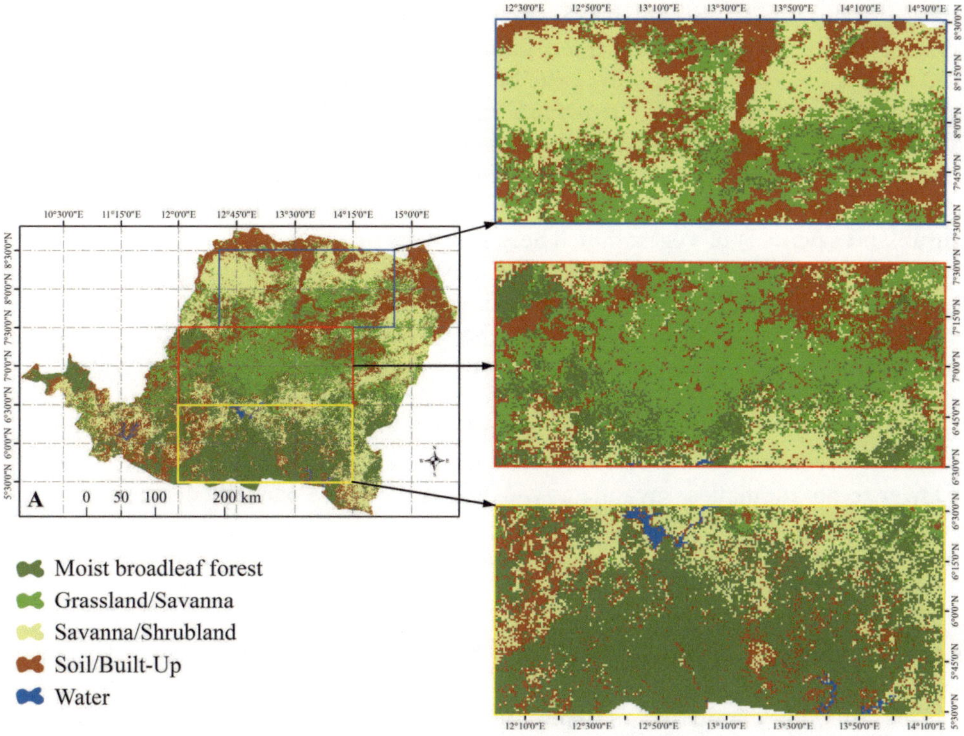

Figure 11.
Zooms on the transitions, a sign of anisotropic distribution with latitudes. Yellow square = lower latitudes (5°30'-6°30'N) transition; red square = middle latitudes (6°30'-7°30'N) transition; blue = upper latitudes (7°30'-8°30'N) transition.

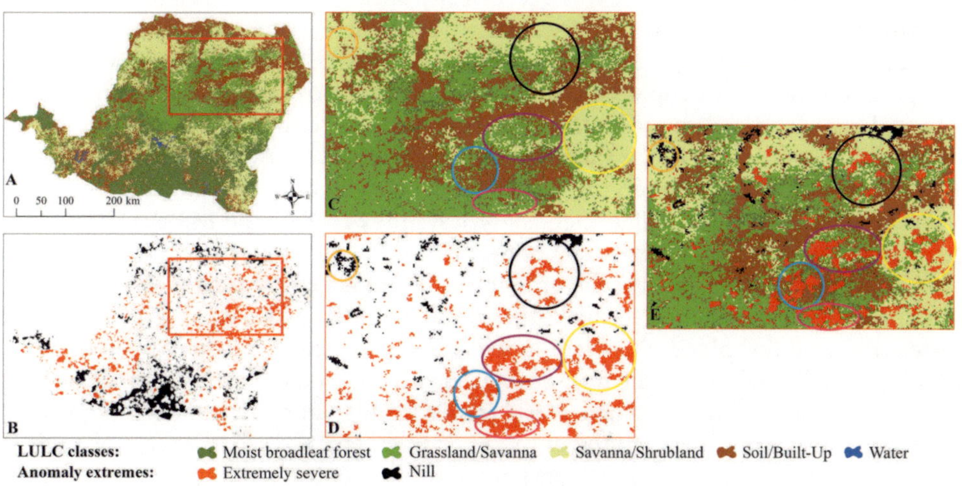

Figure 12.
Anomalies versus LULC classes. (A) SVM classification map. (B) Two extreme classes of anomalies at 3 × 3-pixel moving window size. (C) & (D) subset of comparison among classes of LULC and anomalies. (E) Stacking results with the following details: Black, yellow, purple, and ginger pink circles = extremely severe anomaly spots extended to part of the entire vegetation, in the grassland/savanna –savanna/shrubland transition area; orange circle = low or inexistent anomaly spot in the savanna/shrubland transition, characterized as drier and more exposed vegetation to degradation; cyan circle = low or inexistent anomaly on bare soil.

third ax of vegetation (shrubland savanna), although the second ax (grassland savanna) would have been "expected." Between the middle latitudes $6^0$30'-$7^0$30'North, the transition mixes in the below area, the "-'unexpected" third ax (shrubland savanna) with the 'expected' first ax (moist broadleaf forest) of vegetation, before the wide expansion of the 'expected' second ax (grassland savanna). At this point, the only "expected" transition was inside the upper latitudes $7^0$30'-$8^0$30'North, where the second ax (grassland savanna) gradually gave way to third ax (shrubland savanna) of vegetation. These elements of analysis support the qualifier of "bistable" area, while still questioning the anisotropic distribution with latitudes, and encouraging the finest scale of analysis, i.e., spatial and spectral resolution.

Finally, the display of anomalies with the LULC classes disambiguated the confusion of savanna and degraded forest. The observation was made by overlaying the highest and the lowest values of anomalies in the most concentrate area, on the SVM output. On three spots covered by grassland, shrubland and bare soil, the modeled extremely severe anomalies concern just a part of each class. Whereas, on two spots of lower to no-anomalies, savanna as well as bare soils are partially concerned (**Figure 12**).

5. Conclusions

This study has conducted an experimentation on the forest-savanna vegetation, with the goal of assessing dynamics, assuming anomalies and predicting boundaries. On Google Earth Engine platform and using Sentinel2-A satellite images of seven consecutive dry seasons, from 2015 to 2016 to 2021–2022, twelve spectral indices were selected according to their different phenological and physiological assessment of the vegetation, and other natural features to be discriminate. Using the processing of change vector analysis, CVA, it was successfully observed that each index brings a substantial information, to better assess increase or decrease patterns of the vegetation cover. Further, proposing the averaged moving standard deviation index, aMSDI, to face potential issues of simple MSDI, the scale of spatial trends appraisal was found identical between the same interval **[0–0.083]** for all pixel window sizes, while keeping spatial trends as specific as they are for each selected CVA. As confirmation, all $\boldsymbol{p-values=0, z-scores>}2.5$ there is a high clustering between anomaly pixels, whereas low adjusted R^2 among each analytical index aMSDI and MSAVI2 ones validate the performance of the model. Besides, three main trends of vegetation emerged, i.e., moist broadleaf forest in the south, grassland mixed to savanna in the core and savanna mixed to shrubland in the north, based on CART, RF and SVM classifiers performed on thresholded, PCA regrouped and stacked bands, with **88.7%** OA and **0.86** KC. Finally, taking all the nine aMSDI as entries, a paired cross-correlation mapping helped to identify same general trends for high and low values. Whereas, the application of simple linear combination, SLC, highlighted the important spots of anomalies distribution in the northern part of the subset towards Sahel and desert, but less concentrated in the southern part towards moist forest area. Because the forest-savanna anisotropy with latitudes remains questionable depending on the scale of the study, it can be inferred from the study that, although aMSDI method shows capabilities in semi-dry areas, the choice of indices to use is the responsibility of the author. Whereas, a strong correlation remains to be investigated between the seasonal anomalies and the causes or drivers.

Acknowledgements

To USGS for the free availability of satellite data. To all experts-developers of Google Earth Engine platform APIs. To StatsN'Maps Consulting Firm, for the logistic support. To our laboratories.

Authors contributions

Conception, A.H.N.M.; study design, A.H.N.M.; acquisition of data, A.H.N.M., I.C. N.P., F.C.L.T., L.M.B., M.T. and J.V.M.M.; execution, A.H.N.M.; analysis and interpretation, A.H.N.M.; writing-original draft preparation, A.H.N.M.; writing-review and editing A.H.N.M., I.C.N.P., F.C.L.T., L.M.B., M.T. and J.V.M.M.; All authors have read and agreed to the published version of the manuscript.

Conflict of interest

The authors declare no conflict of interest.

A. Additional material

Links to the code are available upon request.
Additional material 1. Characteristics of Sentinel-2A bands.

	Sentinel-2A MSI	
Name	**Range (μm)**	**Bands (Resolution/m)**
Coastal aerosol	0.421 – 0.457	B1(60)
Blue	0.439 – 0.535	B2(10)
Green	0.537 – 0.582	B3(10)
Red	0.646 – 0.685	B4(10)
Red_edge1	0.694 – 0.714	B5(10)
Red_edge2	0.731 – 0.749	B6(10)
Red_edge3	0.768 – 0.796	B7(10)
NIR_{wide}	0.767 – 0.908	B8(10)
NIR_{narrow}	0.848 – 0.881	B8A(20)
Water vapor	0.931 – 0.958	B9(60)
Cirrus	1.338 – 1.414	B10(60)
SWIR1	1.539 – 1.681	B11(20)
SWIR2	2.072 – 2.312	B12(20)

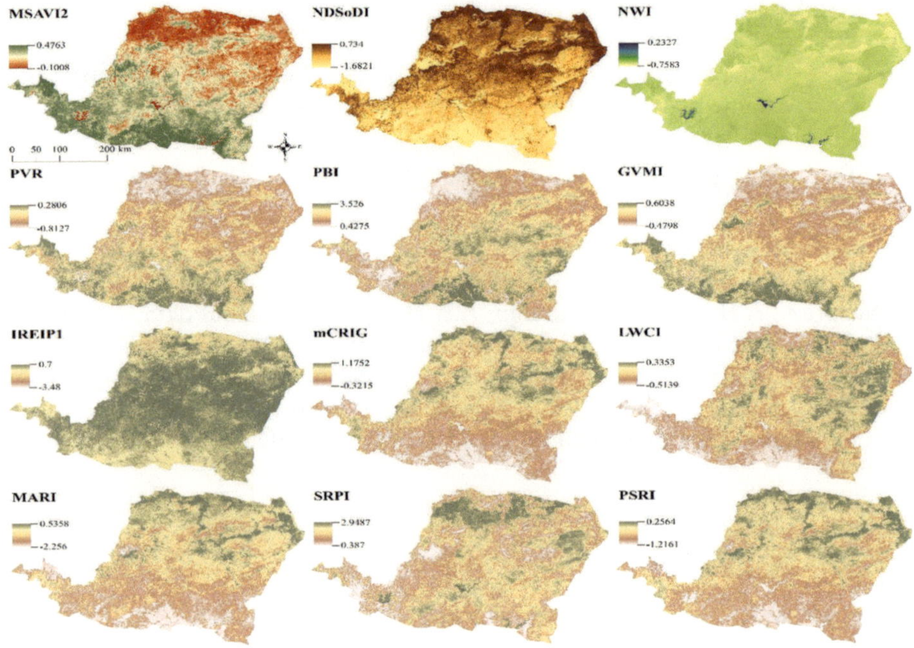

Additional material 2. The spectral indices used. The first row is the supporting or reference data.

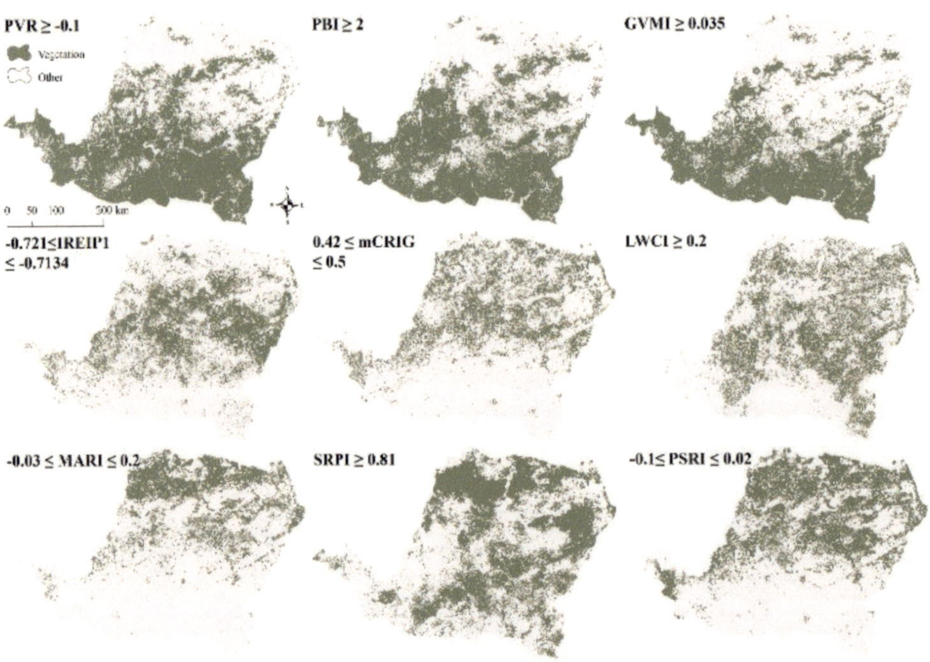

Additional material 3. Binarized indices and thresholds defined.

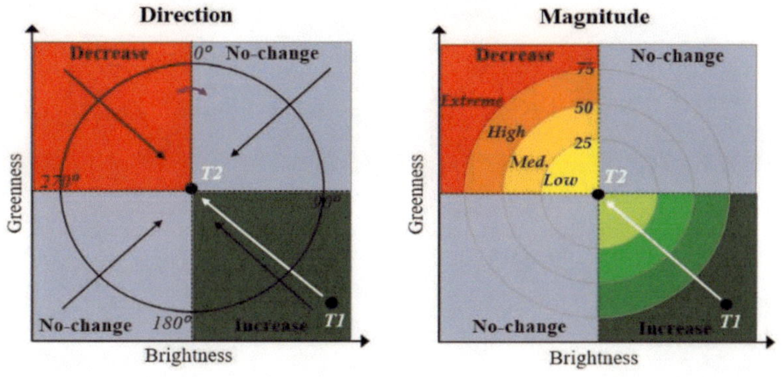

Additional material 4. Direction and magnitude of change as proposed by Kuzera et al. (2005).

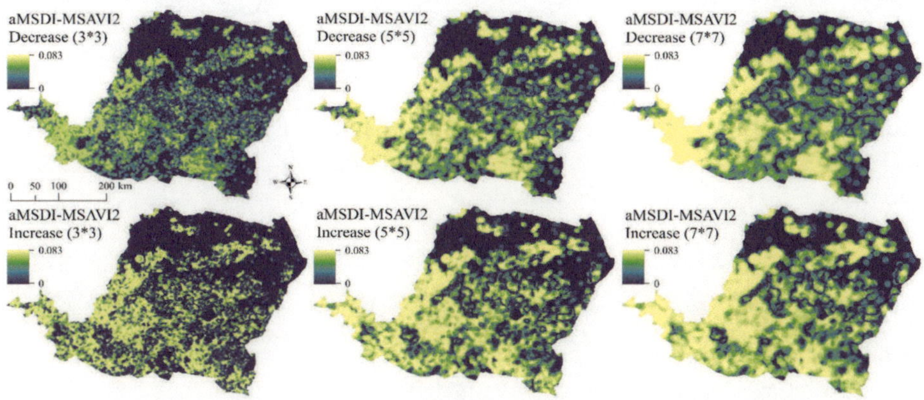

Additional material 5a. Averaged MSDI patterns of MSAVI2.

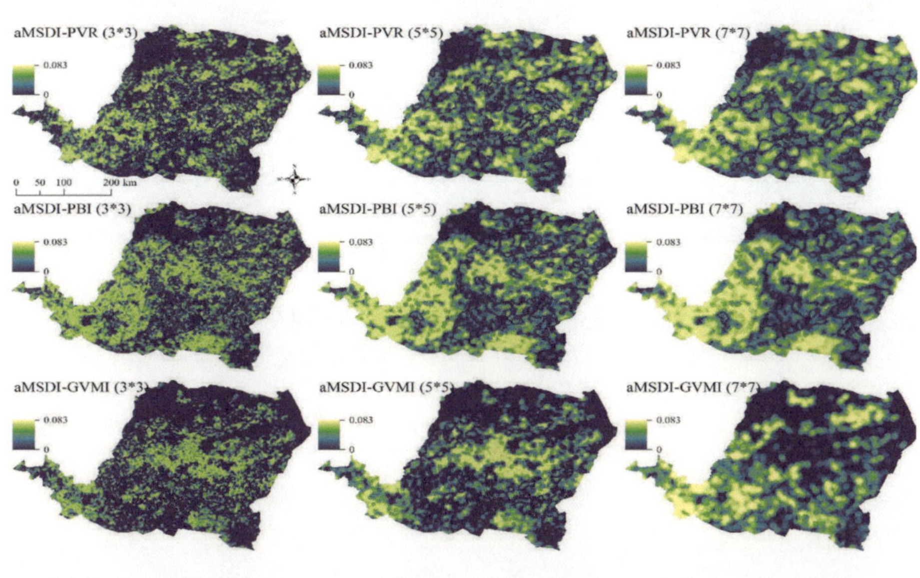

Additional material 5b. Averaged MSDI patterns for PVR, PBI and GVMI.

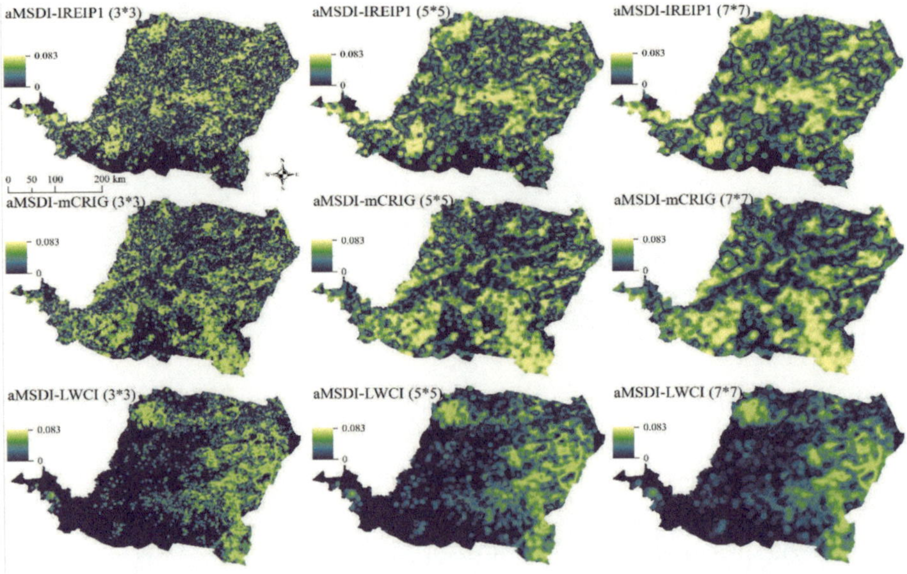

Additional material 5c. Averaged MSDI patterns for IREIP1, mCRIG and LWCI.

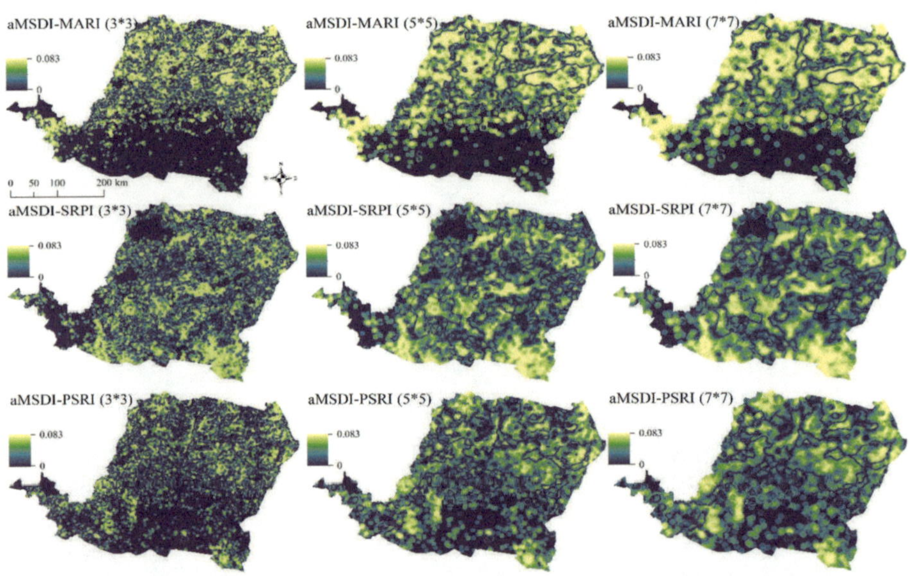

Additional material 5d. Averaged MSDI patterns for MARI, SRPI and PSRI.

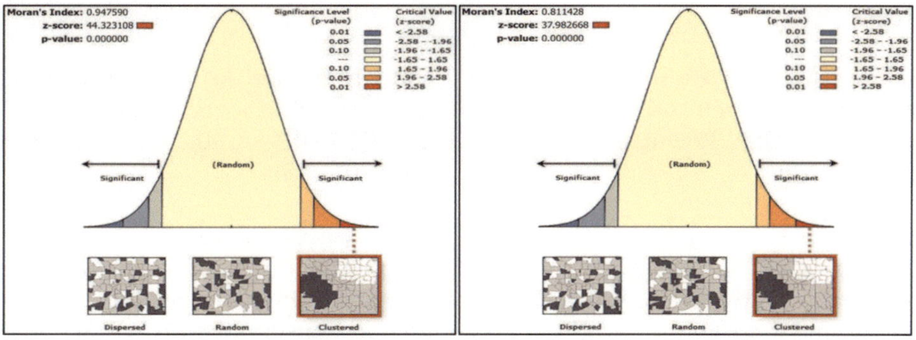

Additional material 6. Spatial autoregression sample plots of aMSDI for PVR (left) and MSAVI2 increase patterns (right). red square = spatial correlation targeted, for n = 1544 pixels.

Author details

Alfred Homère Ngandam Mfondoum[1,2*], Igor Casimir Njombissie Petcheu[2], Frederic Chamberlain Lounang Tchatchouang[2], Luc Moutila Beni[3], Mesmin Tchindjang[2] and Jean Valery Mefire Mfondoum[4]

1 StatsN'Maps, Consulting Firm, Dallas, USA

2 Department of Geography, University of Yaounde, Natural Resources Management Laboratory, Yaounde, Cameroon

3 Department of Geography, University of Douala, Douala, Cameroon

4 Department of Statistics and econometrics, University of Toulouse Capitol 1, Toulouse, France

*Address all correspondence to: stats.n.maps.expertise@gmail.com

References

[1] Johnson DL, Lewis LA. Land Degradation: Creation and Destruction. Lanham, MD, USA: Rowman & Littlefield; 2006

[2] D'Odorico P, Bhattachan A, Davis KF, Ravi S, Runyan CW. Global desertification: Drivers and feedbacks. Advances in Water Resources. 2013;**51**: 326-344. DOI: 10.1016/j.advwatres.2012.01.013

[3] Bullock EL, Woodcock CE, Olofsson P. Monitoring tropical forest degradation using spectral unmixing and Landsat time series analysis. Remote Sensing of Environment. 2018;**238**:1-16. DOI: 10.1016/j.rse.2018.11.011

[4] Chaves MM, Maroco JP, Pereira JS. Understanding plant responses to drought – From genes to the whole plant. Functional Plant Biology. 2003;**30**(3): 239-264. DOI: 10.1071/FP02076

[5] Ratnam J, Bond WJ, Fensham RJ, Hoffmann WA, Archibald S, Lehmann CER, et al. When is a 'forest' a savanna, and why does it matter? Global Ecology and Biogeography. 2011;**20**: 653-660. DOI: 10.1111/j.1466-8238.2010.00634.x

[6] Lehmann CE, Archibald SA, Hoffmann WA, Bond WJ. Deciphering the distribution of the savanna biome. New Phytologyst. 2011;**191**:197-209. DOI: 10.1111/j.1469-8137.2011.03689.x

[7] Staver AC, Archibald S, Levin SA. The global extent and determinants of savanna and forest as alternative biome states. Science. 2011;**334**:230-232. DOI: 10.1126/science.1210465

[8] Aleman JC, Fayolle A, Favier C, Staver AC, Dexter KG, Ryan CM, et al. Floristic evidence for alternative biome states in tropical Africa. Proceedings of the National Academy of Sciences. 2020; **117**(45):28183-28190. DOI: 10.1073/pnas.2011515117

[9] Tucker CJ, Justice CO, Prince SD. Monitoring the grasslands of the Sahel 1984–1985. International Journal of Remote Sensing. 1986;7(11):1571-1581. DOI: 10.1080/01431168608948954

[10] Townshend JRG, Justice CO. Analysis of the dynamics of African vegetation using the normalized difference vegetation index. International Journal of Remote Sensing. 1986;7(11):1435-1445. DOI: 10.1080/01431168608948946

[11] Anchang JY, Prihodko L, Ji W, Kumar SS, Ross CW, Yu Q, et al. Toward operational mapping of Woody canopy cover in tropical savannas using Google earth engine. Frontiers of Environmental Science. 2020;**8**(4):1-17. DOI: 10.3389/fenvs.2020.00004

[12] Hansen MC, Roy DP, Lindquist E, Adusei B, Justice CO, Altstatt A. A method for integrating MODIS and Landsat data for systematic monitoring of forest cover and change in the Congo Basin. Remote Sensing of Environment. 2008;**112**(5):2495-2513. DOI: 10.1016/j.rse.2007.11.012

[13] Ngandam Mfondoum AH, Etouna J, Buji KN, Mvogo Moto FA, Noulaquape Deussieu FG. Assessment of land degradation status and its impact in arid and semi-arid areas by correlating spectral and principal component analysis neo-bands. International Journal of Advanced Remote Sensing and GIS. 2016;**5**(2): 1539-1560. DOI: 10.23953/cloud.ijarsg.77

[14] Ngandam Mfondoum AH, Gbetkom PG, Cooper R, Hakdaoui S, Mansour Badamassi MB. Improving the

land surface general drought index model. The International Archives of the Photogrammetry, Remote Sensing and Spatial Information Sciences. 2020;**XLII-3/W11**:101-108. DOI: 10.5194/isprs-archives-XLII-3-W11-101-2020

[15] Gadal S, Gbetkom PG, Ngandam Mfondoum AH. A new soil degradation method analysis by sentinel 2 images combining spectral indices and statistics analysis: Application to the Cameroonians shores of Lake Chad and its hinterland. Proceedings of the 7th International Conference on Geographical Information Systems Theory, Applications and Management. 2021;25-36. DOI: 10.5220/0010521200250036

[16] Karnieli A, Kaufman YJ, Remer L, Wald A. AFRI – Aerosol free vegetation index. Remote Sensing of Environment. 2001;**77**:10-21. DOI: 10.1016/S0034-4257(01)00190-0

[17] Rouse JW, Haas RH, Schell JA, Deering DW, Harlan JC. Monitoring the Vernal Advancements and Retrogradation (Greenwave Effect) of Nature Vegetation. Greenbelt, MD, USA: NASA, NASA/GSFC Final Report; 1974

[18] Huete AR, Liu H, Van Leeuwen WJD. The use of vegetation indices in forested regions: Issues of linearity and saturation. IEEE International Geoscience and Remote Sensing Symposium Proceedings. 1997;**4**:1966-1968. DOI: 10.1109/IGARSS.1997.609169

[19] Ghulam A, Qin Q, Teyip T, Li Z-L. Modified perpendicular drought index (MPDI): A real-time drought monitoring method. ISPRS Journal of Photogrammetry and Remote Sensing. 2007;**62**(2):150-164. DOI: 10.1016/j.isprsjprs.2007.03.002

[20] Ali AM, Darvishzadeh R, Skidmore AK. Retrieval of specific leaf area from Landsat-8 surface reflectance data using statistical and physical models. IEEE Journal of Selected Topics in Applied Earth Observations and Remote Sensing. 2017;**10**(8):3529-3536. DOI: 10.1016/j.agrformet.2017.01.015

[21] Tsalyuk M, Kelly M, Getz WM. Improving the prediction of African savanna vegetation variables using time series of MODIS products. ISPRS Journal of Photogrammetry and Remote Sensing. 2017;**131**:77-91. DOI: 10.1016/j.isprsjprs.2017.07.012

[22] Olson DM, Dinerstein E, Wikramanayake ED, Burgess ND, Powell GV, Underwood EC, et al. Terrestrial ecoregions of the world: A new map of life on earth: A new global map of terrestrial ecoregions provides an innovative tool for conserving biodiversity. Bioscience. 2001;**51**(11):933-938. DOI: 10.1641/0006-3568(2001)051[0933:TEOTWA]2.0.CO;2

[23] Baillarin SJ, Meygret A, Dechoz C, Petrucci B, Lacherade S, Tremas T, et al. Sentinel-2 level 1 products and image processing performances. IEEE International Geoscience and Remote Sensing. 2012;**39**(B1):197-202. DOI: 10.1109/IGARSS.2012.6351959

[24] Gorelick N, Hancher M, Dixon M, Ilyushchenko S, Thau D, Moore R. Google earth engine: Planetary-scale geospatial analysis for everyone. Remote Sensing Environment. 2017;**202**:18-27. DOI: 10.1016/j.rse.2017.06.031

[25] Scanlon TM, Caylor KK, Manfreda S, Levin SA, Rodriguez-Iturbe I. Dynamic response of grass cover to rainfall variability: Implications for the function and persistence of savanna ecosystems. Advances in Water Resources. 2005;**28**(3):291-302. DOI: 10.1016/j.advwatres.2004.10.014

[26] Chen J, Menges C, Leblanc S. Global mapping of foliage clumping index using multi-angular satellite data. Remote Sensing of Environment. 2005;**97**(4): 447-457. DOI: 10.1016/j.rse.2005.05.003

[27] Qi J, Chehbouni A, Huete AR, Kerr YH, Sorooshian S. A modified soil adjusted vegetation index. Remote Sensing of Environment. 1994;**48**:119-126. DOI: 10.1016/0034-4257(94)90134-1

[28] Ding F. A new method for fast information extraction of water bodies using remotely sensed data. Remote Sensing Technology and Application. 2009;**24**(2):167-171. DOI: 10.11873/j.issn.1004-0323.2009.2.167

[29] Metternicht G. Vegetation indices derived from high-resolution airborne videography for precision crop management. International Journal of Remote Sensing. 2003;**24**(14):2855-2877. DOI: 10.1080/01431160210163074

[30] Ceccato P, Gobron N, Flasse S, Pinty B, Tarantola S. Designing a spectral index to estimate vegetation water content from remote sensing data: Part 1: Theoretical approach. Remote Sensing of Environment. 2002;**82**(2–3):188-197. DOI: 10.1016/S0034-4257(02)00037-8

[31] Index Database: IDB - N^{0}378. https://www.indexdatabase.de/db/i-single.php?id=212

[32] Vogelmann GE, Rock BN, Moss DM. Red edge spectral measurements from sugar maple leaves. International Journal of Remote Sensing. 1993;**14**(8): 1563-1575. DOI: 10.1080/01431169308953986

[33] Gitelson AA, Keydan GP, Merzlyak MN. Three-band model for noninvasive estimation of chlorophyll, carotenoids, and anthocyanin contents in higher plant leaves. Geophysical Research Letters. 2006;**33**(11):1-5. DOI: 10.1029/2006GL026457

[34] Hunt ER, Rock BN, Nobel PS. Measurement of leaf relative water content by infrared reflectance. Remote Sensing of Environment. 1987;**22**(3): 429-435. DOI: 10.1016/0034-4257(87)90094-0

[35] Main R, Azong Cho M, Mathieu R, O'Kennedy MM, Ramoeloa A, Koch S. An investigation into robust spectral indices for leaf chlorophyll estimation. ISPRS Journal of Photogrammetry and Remote Sensing. 2011;**66**(6):751-761. DOI: 10.1016/j.isprsjprs.2011.08.001

[36] Peñuelas J, Gamon JA, Fredeen AL, Merino J, Field CB. Reflectance indices associated with physiological changes in nitrogen- and water-limited sunflower leaves. Remote Sensing of Environment. 1994;**48**(2):135-146. DOI: 10.1016/0034-4257(94)90136-8

[37] Merzlyak MN, Gitelson AA, Chivkunova OB, Rakitin VY. Non-destructive optical detection of pigment changes during leaf senescence and fruit ripening. Physiologya Plantarum. 1999; **106**(1):135-141. DOI: 10.1034/j.1399-3054.1999.106119.x

[38] Lambin EF, Strahler AH. Change-vector analysis in multitemporal space: A tool to detect and categorize land-cover change processes using high temporal-resolution satellite data. Remote Sensing of Environment. 1994;**48**:231-244. DOI: 10.1016/0034-4257(94)90144-9

[39] Kuzera K, Rogan J, Eastment JR. Monitoring vegetation regeneration and deforestation using change vector analysis: mount St. Helens study area. ASPRS Annual Conference. 2005; DOI: https://citeseerx.ist.psu.edu/viewdoc/download?doi=10.1.1.546.1891&rep=rep1&type=pdf

[40] Tanser FC, Palmer AR. The application of a remotely-sensed diversity index to monitor degradation patterns in a semi-arid heterogeneous south African landscape. Journal of Arid Environments. 1999;**43**(4):477-484. DOI: 10.1006/jare.1999.0568

[41] Schlesinger WH, Reynolds JF, Cunningham GL, Huenneke LF, Jarrell WM, Virginia RA, et al. Biological feedbacks in global desertification. Science. 1990;**247**(4946):1043-1048. DOI: 10.1126/science.247.4946.1043

[42] Robinson JA, Lulla KP, Kashiwagi M, Suzuki M, Nellis MD, Bussing CE, et al. Conservation applications of astronaut photographs of earth: Tidal-flat loss (Japan), elephant effects on vegetation (Botswana), and seagrass and mangrove monitoring (Australia). Conservation Biology. 2001;**15**:876-884. DOI: 10.1046/j.1523-1739.2001.015004876.x

[43] Fullman TJ, Bunting EL. Analyzing vegetation change in an elephant-impacted landscape using the moving standard deviation index. Land. 2014;**3**:74-104. DOI: 10.3390/land3010074

[44] Pickup G, Chewings VH. Forecasting patterns of erosion in arid lands from Landsat MSS data. International Journal of Remote Sensing. 1988;**9**(1):69-84. DOI: 10.1080/01431168808954837

[45] Cliff AD, Ord JK. Spatial Autocorrelation. London, UK: Pion Limited; 1973

[46] Brunsdon C, Fotheringham AS, Charlton ME. Geographically weighted regression: A method for exploring spatial nonstationarity. Geographical Analysis. 1996;**28**(4): 281-298. DOI: 10.1111/j.1538-4632.1996.tb00936.x

[47] Breiman L, Friedman JH, Olshen RA, Stone CJ. Classification and Regression Trees. 1st ed. UK: Routledge; 1984

[48] Breiman L. Random forests. Machine Learning. 2001;**45**:5-32

[49] Mountrakis G, Im J. Ogole C support vector machines in remote sensing: A review. ISPRS Journal of Photogrammetry and Remote Sensing. 2011;**66**(3):247-259. DOI: 10.1016/j.isprsjprs.2010.11.001

[50] Congalton RG, Green K. Assessing the Accuracy of Remotely Sensed Data—Principles and Practices. 2nd ed. Boca Raton, FL, USA: CRC Press; 2009

[51] Landis JR, Koch GG. A one-way components of variance model for categorical data. Biometrics. 1977;**33**(4): 671-679. DOI: 10.2307/2529465

[52] Jafari R, Lewis MM, Ostendorf B. An image-based diversity index for assessing land degradation in an arid environment in South Australia. Journal of Arid Environments. 2008;**72**: 1282-1293. DOI: 10.1016/j.jaridenv.2008.02.011

Chapter 5

The Metropolitan Transformation of Ioannina City from 1940 to 2015

Efthymios-Spyridon Georgiou

Abstract

The chapter presents the urban and regional changes in the city of Ioannina, Greece. This city is located in the periphery of Epirus, which is in the western Balkans, Eastern Europe. The chapter examines, with the tools of aerial photos and QGIS software, the spatial transformation of Ioannina city from 1940 to 2015. Map science is a field through which the users could observe and compare maps from past to future. The plans and the planning were formed under the values, standards, and fundamentals of the mosaic of politics, good practices, urban rules, and citizen level. The urban space has already changed until nowadays. The chapter examines the reasons for urban politics and social–economic moments that became the epitome of these urban and regional changes. The results show the comparative spatial study from each historical period.

Keywords: analysis, cartography, urbanism, policies, Ioannina city

1. Introduction

This chapter is an independent effort to analyze and document the urban transformation in Ioannina, Epirus, Greece. The metropolitan area of Ioannina is a mountainous area, located in northern Greece and the western Balkans. It is an area of rich natural and environmental wealth. The main advantages are the cultural and architectural heritage, its important history, and the natural beauty of the wider metropolitan area. The analysis of the metropolitan area, from 1940 to 2015, presents characteristics of change, innovation, modernism, modernization, and Europeanism. At the same time, the traditional and historical characters have not only been preserved but also have been restored, maintained, and adapted to the modern requirements of everyday life. Finally, the functional character of the main and secondary roads is particularly significant, because they serve a key role in hiking and recreational activities for young and older citizens.

2. Methodology steps

The methodology was based on the creation, analysis, classification, and hierarchy of the metropolitan and urban areas of Ioannina. Initially, the emphasis on detail was essential because the analysis parts were sorted into pieces, and the primary spatial data were prioritized. Also, open-source software for geographic information systems was used to create the spatial and descriptive data of the Balkan cities. The result of the mapping was presented with images and paintings. Then, schematically, the data analysis methodology was shown concisely, intelligibly, and effectively. In summary, in the end, the conclusions of the research and the results from the spatial, cultural, political, geographical, historical, and demographic study were presented.

2.1 The basic policies of urban transformation

This chapter presents the key historical events and pillars of urban change from 1940 to 2015. The end of World War II was the first beginning of urban reconstruction. The urban and spatial changes that took place in the Balkan region are related to the general social and economic developments in the Ottoman Empire as well as the nationalist aspirations of the Balkan countries. The appearance of capitalism in the regions was associated with the development of a new urban environment [1].

2.1.1 The continuation of the war in the region of Pindos

The resistance struggle of Greeks, with the slogan "NO," did not end after the occupation by the invaders in the Balkan region. At that time, the educational level of Greeks was shallow. Few people knew how to write and were highly educated. This fact is added to the high rural population (over 80%) of the Greek territory living in rural, mountainous, and nonurban areas. The world was affected and divided because of poverty.

2.1.2 The Marshall plan

American economic aid was the first firm policy since the end of the war. The Marshall Plan was intended to solve manual work and low educational attainment. It was a key pillar of urbanization, as schools, universities, squares, residences and roads were created, squares in urban centers. Of course, this financial aid did not change the situation from one day to the next, but it lasted for decades. In summary, it was the beginning of liberal democracy and the formation of the everyday routine. With Athens as the central pole, housing and education developed. Since 1974, the democratization of the country has followed. It is essential that at that time in Ioannina, five main trends were formed:

- The residents of the metropolitan area came to study in Ioannina.
- Trade was developed. The basic traditional products of the settlements for domestic, friendly, related use were transformed into the first shops, guilds, and cooperatives.
- The wealthy and historic families moved to Athens and abroad. There was also a lot of emigration from very poor families. The first road axes began to be constructed with the benefits of wealthy Epirus's. For example, the roads of Dodoni and E. Averoff.

- Emphasis was placed on preserving the forces of the past, such as on the principles, values, and visions of the Greek revolution. For example, in Ioannina of the late Ottoman Empire, there was the local expression that the city was "first in arts, wealth, and letters."

2.1.3 The era of transition and integration in the European Union

A significant change occurred in 1990, with the fall of the socialist-communist regimes. At this point, models of liberal democracy in the city peaked. The inhabitants of the mountainous areas of northern Pindos contributed to the rapid increase of the reconstruction of the metropolitan area of Ioannina. The houses were built, and the first settlements in the wider area were established. These settlements in the following years took the form of small towns with new villas.

On May 28, 1979, the treaty of accession of Greece to the European communities was signed at the Zappeion Megaron. Greece formally joined the European Communities on 1 January 1981 and became its tenth member [2]. The entrance of E.U. gave an essential boost to democracy, liberalism, the opening of borders, and the development of small businesses (bakeries, boutiques, and grocery stores). Moreover, a key aspect was the emigration of people from communist-socialist Albania, who had of Greek-speaking origin, before the borders were closed. These people worked as stone craftsmen and specialized in concrete structures. The reconstruction in the metropolitan area of Ioannina was mainly based on these inhabitants. The residents worked in neighboring regions and cities, such as Thessaloniki, with higher salaries and better working positions.

2.1.4 The historical background of construction

Nowadays, the developmental physiognomy of Ioannina has a basis in the flourishing of tourist products, environmental and ecological activities, recreation, education, business, and administrative services.

2.2 The developmental physiognomy of the Ioannina nowadays

Today, Ioannina is the largest city in north-western Greece and one of the largest cities in the Balkan Peninsula in population. The region of Epirus is one of the 13 regions of Greece. Administratively, it is an appellate body of local government. The geographically is identified with the homonymous district. It occupies an area of 9203 sq. km. Furthermore, its population amounts to 336,856 inhabitants.

2.2.1 The metropolitan role

With this innovative, modern, and efficient tool, the Ioannina basin acquires excellent benefits because it is the largest administrative center in the southern-western Balkans. The role and importance are constantly in a strong position because of its historical and cultural background, education, business, and administrative services.

Figure 1a and **b** show the region of Ioannina in Greece. Epirus covers the west part of Greece. Ioannina is located near Italy, Albania and Patra. Nowadays, the crucial roads connect from the port of Igoumenitsa to Istanbul and the road from Ioannina to Athens. To sum up, the strategic position is the main advantage of the city.

(a)

(b)

Figure 1.
(a) Epirus and Ioannina and (b) Ioannina in Greece.

Municipality	403.32 km^2 (155.72 sq. mi)
Municipal unit	47.44 km^2 (18.32 sq. mi)
Elevation	480 m (1,570 ft)
Municipality density	280/km^2 (720/sq. mi)
Municipal unit	80,371
Municipal unit density	1700/km^2 (4400/sq. mi)
Municipality	112,486

Table 1.
The Municipality in Ioannina.

Table 1 presents the geomorphology and demographic characteristics of the municipality in Ioannina. For example, the elevation is 480 meters and the area of the municipality is 403.32 km^2. Also, the table preserves the biggest cities in the west Balkans. The biggest city is Beograd in Serbia and the second is Zagreb. The city of Ioannina has 112.486 inhabitants and the municipality has 336.856 inhabitants.

2.2.2 Mapping in QGIS

Figures 2 and **3** and **Table 2** show the largest population cities in the western Balkans. The map was constructed with QGIS. The methodology was based on the addition of the Google Satellite background. The shape files of cities were designed. In addition, the spatial and descriptive archive of towns was created. The cities have fields in the illustrative data table, name and population.

With the editing process, the points were formed, the size was increased, and the color was modified. Finally, with help of the labeling toolbox, the field name was activated, and the names of the cities appeared on the map. The map was exported as an image in scale, representing all the cities of the western Balkans [3].

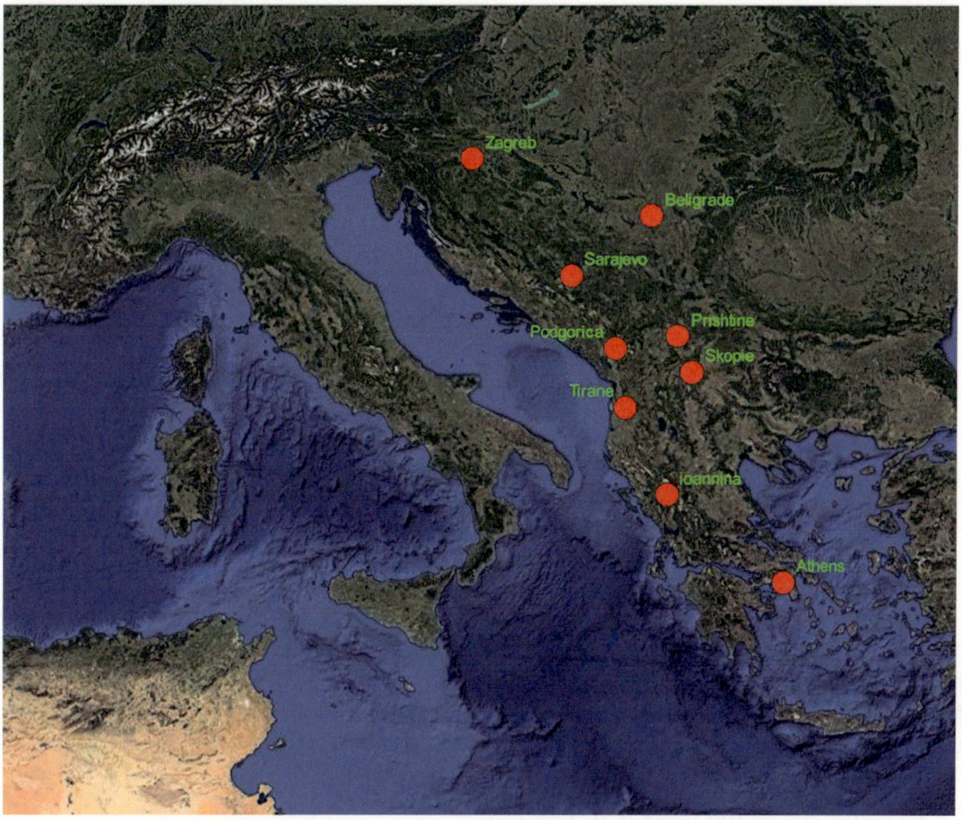

Figure 2.
Metropolitan area of Ioannina.

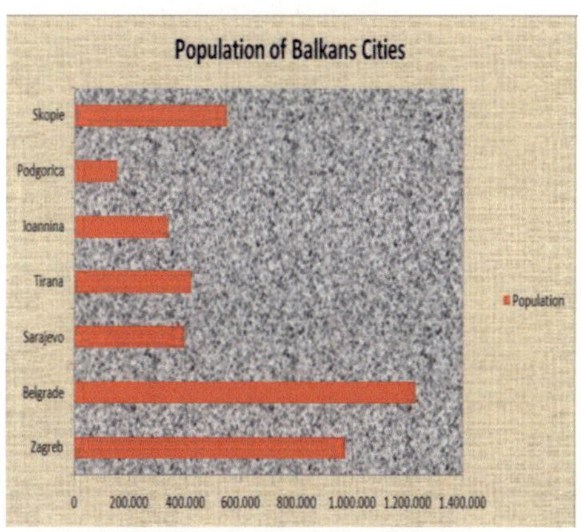

Figure 3.
The population of Balkans cities.

New technology for assessing the technical condition of datasets and vector sets is proposed. The core of this technology is an estimation of defects and deformations of structures using QGIS during map construction and operation [4].

Cities	Population
Zagreb	973.667
Belgrade	1.230.000
Sarajevo	395.133
Tirana	418.495
Ioannina	336.856
Podgorica	156.000
Skopje	546,824

Table 2.
The biggest cities in the west Balkans.

The methodology is based on data management analysis and QGIS software. Firstly, the author searches the pertinent data about the population in the Balkans cities and demographic sizes. Next, the author creates diagrams with the use of excel office. The diagrams present an innovative understanding of the population, municipality area, elevation, and destiny.

Also, this data uses in QGIS. The following steps create a point layer relative to the name of the cities. This layer shows the main cities in the Balkan area and the classification is about the size of the population. Finally, the user presents the methodology of the progress in the diagram. The progress in one part of the scientific study of spatial analysis assistance of the QGIS (**Figure 4**).

2.3 The urban transformation in 1940

Figure 5a and **b** show Ioannina in the year 1940. The urban fabric of Ioannina extends around the lake of Ioannina, which is called Pamvotis. In the figure, there are 4–5 main arteries, which are connected together. There are huge and small roads, which are given below.

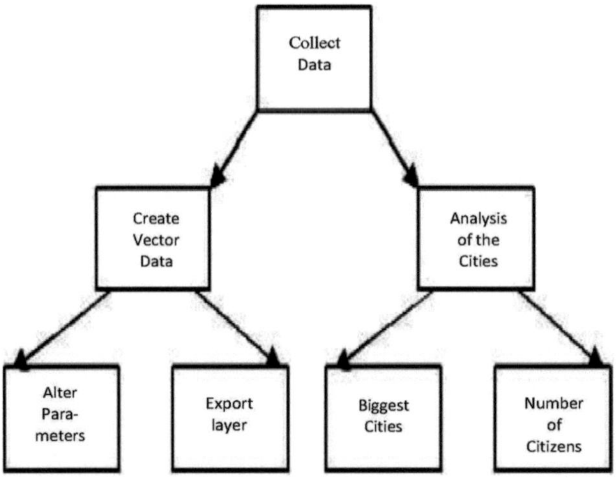

Figure 4.
The methodology in Data Management.

Figure 5.
(a) Aerial photo from Ioannina in 1940 and (b) Aerial photo from Ioannina in 1940.

- The lakeside road.
- The road parallel to the main lakeside artery.
- The road that connects the city with the west entrance.
- The road that connects the city with the eastern exit.
- The north entrance, in the grove of Frontzos.

The urban density is concentrated in a rectangular shape, extending from Mavili Square perpendicular to the park of Frontzos. On both sides of the city, on the west and east, there are large forest areas and bushy cultivation.

The islet of Ioannina is in the lakeside zone of the city. An essential history from the "palingenesis" of the Greek statement. The small island travel is enough to cross it by small boat in about 20-25 minutes. In the old days, many people lived on the island because they were fishermen, merchants, and small - owners of traditional and folk art. The island is famous for its local cuisine, such as the frog's legs.

2.3.1 Coastal front

The coastal front during this period was a place for swimming, fishing, and recreation. The lake "gave" work to the island's inhabitants and the castle. The inhabitants were among the most economically robust in the city. In addition, there were significant flora and fauna around the lake. Finally, the lake was an important historical, cultural, architectural, and political place to re-create the Greek state. The reasons are important events, battles, and political decisions. The site preserves the historical tomb with museums and cultural-historical sites.

2.3.2 Road network

The road network in this historical period is not safe and healthy. The effects of the second war are influential in the city. The main roads were built with money from donations. The wealthy Epirus persons give money to construct parks, roads, and squares. Moreover, the city has an essential connection to the entrances of the Ioannina. For example, the steep road, which connects the lakeside zone with the city's eastern entrance, was the best–maintained. On this street was the historic square of Ioannina, the old clock, the academy, the historic town hall, and the main commercial activity of the residents. Finally, Ioannina is rich in stone reserves because mountains surround the city.

2.3.3 Green areas

Ioannina is an area with a large percentage of greenery. The more comprehensive region benefits from the geomorphology of mountains, rivers, lakes, and forests. In 1940, there was no urbanization in Ioannina. In fact that it was an advantage in the agricultural and livestock sector. Agricultural land was mostly for animal husbandry, while fields were necessary for fodder feed.

The central and most fundamental part of economic life was based on this triptych. The dairy economic function is the cultivation - of sheep - dairy products. Finally, the inhabitants had small crops with vegetables and fruit, contributing to a healthy lifestyle and self-sufficiency.

2.3.4 Main economic activities

In the metropolitan area, the main activities of inhabitants are animals and agriculture. The most significant percentage of the population lived in settlements in mountainous, semi-mountainous, and lowland volumes.

The inhabitants have the purpose of being self-sufficient. For example, the sheep and goats produced milk, cheese, and yogurt. Also, the gardens and fields gave vegetables, fruits, potatoes, and cereals. In this way, the residents made the necessary biological needs for food and sure profits, which sold the products that were left over.

Ioannina was a large administrative center of the Ottoman Empire. There were small - ownership, folk art, technical professions, fish shops, patisseries, bakeries, and professional cooperatives.

2.4 The land uses changes in 2007

The map below depicts the urban fabric of Ioannina in 2007. Ioannina, at this time, is a city that has been transformed into a modern educational, university, sports, and administrative center in Greece.

There is a gradual transformation as a living cell of culture, democracy, and liberal economy, where the forces of conservation strengthen the developmental physiognomy of Ioannina. The city was an important pole of the economic health in Greece during the period 2000–2010.

2.4.1 The Egnatia highroad

An important role is played by the dispute of Egnatia Odos, a landmark road project for the region. The main benefits were the reduction of the equation in time - distance with neighboring regions and the growth of tourist flows from Western Europe and the Balkans.

This was a turning point in the city's modern history, which began to augmentation the leading role in the region. Today, it is a road junction connecting the Egnatia and Ionian highroads (**Figure 6**).

Figure 6.
(a) Aerial photo from Ioannina in 2007 and (b) aerial photo from Ioannina in 2007.

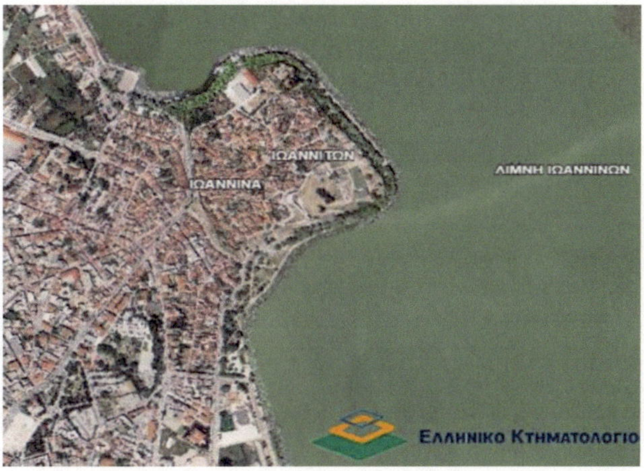

Figure 7.
Aerial photo from Ioannina in 2007.

This photo is the culmination of the increase in the urban complex of Ioannina. The policies of previous years were crystallized with images. The increase in migratory flows, sustainable development, and improvement of the level of education were crucial pillars of the development process of Ioannina.

The main differences are that now the houses have roofs with ceramic bricks, and the roads' quality and quantity have improved. The secondary and local roads have increased, and many maintenance and restoration works have been carried out in the old mansions.

In addition, the coastal road and widening the pedestrian street linearly with the lake is essential. Finally, small "hearths" of greenery have been developed within the city's urban fabric (**Figure 7**).

2.4.2 European policies

During this period, the city passed a phase of modernization. The economic, cultural, and tourist developmental activities harmonized with E.U political directions. Ioannina became a cell of democracy, neoliberal politics, a tourist attraction, environmental and cultural center. The main pillars of development are around (**Figure 8**).

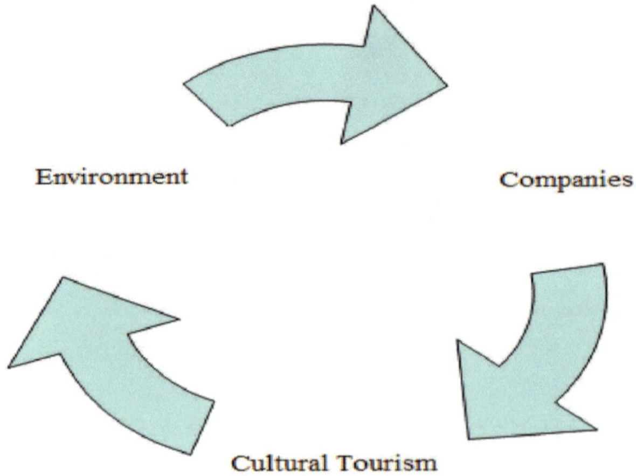

Figure 8.
The triptych of economic development.

2.4.3 Coastal front

The lakeside front of Ioannina is a pole of development. The lake gathers economic activities. In the old days, the island's inhabitants and the castle area were fishermen and landowners. Today, tourist activities, folk art, cafes, museums, restaurants, and hotels have been developed. The lakeside front of Ioannina is an essential pillar for increasing the per capita income of the inhabitants.

Green fields, parks, playgrounds, and event spaces are gathered linearly. Especially during the summer, there are tourist flows, good entertainment, sports competitions, concerts, and sweet and agricultural products exhibitions. Finally, the factor is important because the cultural character around the lake is an opportunity for more development. It also contributes to the sustainable development of Ioannina. After all, it is the historic center of Ioannina.

2.4.4 Metropolitan benefits

The urban fabric of the city consists of structure and urbanism of the following elements:

1. Three main areas

    - Square Pargis

    - Prefecture Square—Market

    - Square Mavili—Lakeside Front

2. Three main entrances

    - North Entrance—Front Frontzos

DOI: http://dx.doi.org/10.5772/intechopen.105884

- Airport Entrance
- Entrance Gianniotikos Salon

3. Urban areas

- Historical center
- Spaces Pure Residences—Settlements
- Administration reas
- Restaurants
- Green Spaces—Hiking areas, hotels, – spaces for events and recreation

4. Traditional settlements

   A significant benefit is a proximity to traditional settlements of particular natural beauty. The advantages of the mountainous and semi-mountainous settlements are the traditional architecture, the restaurants, the hotel units, the cuisine, and the activities in the natural area.

The metropolitan area of Ioannina is a geographical area of Greece with significant advantages. The geographical position and geomorphology include access to the sea and mountains. Also, the historicity, the cultural heritage, and the natural environment are the strong points of the basin of Ioannina. The developmental physiognomy is based on tourist flows. This politic adds economic value to the residents of the wider region. For example, the proximity to Italy from the air and road connection with Athens is a straightforward solution for tourists.

Also, the role of Ioannina has always been an administrative center and is essential. The rich history confirms this. Today, the responsibilities have been significantly upgraded with the upgrading of the university, the region of Ioannina, the university hospital, the hotels, the services of the municipality, and industries.

3. The metropolitan area of Ioannina

In 2015, Ioannina was upgraded for two main reasons. Initially, it is now a crossroads of two major roads of the country. The Ionian and Egnatia Odos are joined 9 km from the city. The Ionian Sea connects the town with Patras and Athens, while Egnatia connects Ioannina with the Ionian Sea, Thessaloniki, and Istanbul. Also, the city is a cultural, historical, environmental, tourist, and administrative center of Greece. The increase in tourist flows through neighboring Italy and other Balkan towns contributed to the rise of visitors.

The unique natural beauty areas of the neighboring mountains are a mixture of the environment, natural beauty, and tradition. Also, the benefits of the traditional mountain settlements are significant. The buildings of the settlements were

Figure 9.
The urban area of Ioannina.

Figure 10.
The historical center.

Figure 11.
The metropolitan area of Ioannina.

preserved. The main factors in the development of the wider region are the churches, monuments, restaurants, large hotel units, hiking trails, lakes, and traditional celebrations.

Figure 12.
The shape of the city in 2007.

Finally, **Figures 9** and **10** are two images of the urban fabric and the historic center of Ioannina. The historic center has been preserved and modernized. Also, the city's urban fabric shows an increase in the west and east pieces (**Figure 11**).

The following images show the city in 2007 and 2015. In **Figure 12**, the urban fabric of the city is gray. **Figure 13** shows the shape of the urban fabric in gray—the green color of the forest cradles and parks. The urban fabric's urban density, perimeter, and area increased from 2007 to 2015. Cartography is usually carried out during building design, planning, and construction. The map contains a variety of important information [5].

In the end, the mapping process is a tremendous spatial example. The analysis of data, topological information, attribute information, and overlay study explores scenarios for the practical knowledge of the perimeters, area, and land covers of Ioannina.

The methodology of spatial representation in the city's shape became the QGIS software. Initially, the images were added to the spatial surface. Also, the shapefiles were created in 2007 and 2015. In continuous, digitization began with the editing tool of the city. Each shapefile was formed based on the urban fabric in the two chronological periods on the image. In the field of descriptive characteristics, the area, the perimeter of the urban fabric, and vegetation were added [3].

4. Conclusion

The message of the scientific paper is concluded in the three words: thesis, antithesis, and synthesis. Firstly, the idea, because the whole article included:

- Basic points,

- Pillars,

Figure 13.
The shape of the city in 2015.

- Fundamentals of the political decisions,

- Urban and metropolitan transformations.

Secondly, antithesis, why urbanism changed during the chronological period from 1940 to 2015, and the land uses reshaped in other forms. This anti-phase is crucial to studies as a human body that changes yearly. The years keep the town together, harmonizing with European values and standards. In other words, it is the same, and the city is a little planet of democracy and theoretical and practical ideas.

Thirdly, the word synthesis. The political decisions continue the cohesion policies. The next political steps follow the previous goals of the European Union. The mosaic of these changes synthesizes the modern reality in the city.

Acknowledgements

I would like to express my thanks to Paula Gavran because she allowed me to participate in this study and write an article for a scientific journal. I am thankful for her suggestion and pleased and enthusiastic about her cooperation. It is an honor for me. I wish her every success.

Author details

Efthymios-Spyridon Georgiou
Freelancer Urban Planner, Athens, Greece

*Address all correspondence to: efthimios_georgiou@yahoo.gr

References

[1] Georgiou ES. City History, Si Magazin. 2020. Available from: https://www.simagazin.com/en/si-urban-en/topics-urban/urban/stadtgeschichte/

[2] Official Website in E.U. 2022. Available from: https://greece.representation.ec.europa.eu/shetika-me-emas/i-ellada-stin-ee_el. [Accessed May 7, 2022]

[3] Official Website QGIS Training Manual. 2022. Available from: https://docs.qgis.org/3.22/en/docs/training_manual/. [Accessed: May 7, 2022]

[4] Annenkov A. Monitoring the deformation process of engineering structures using bim technologies. International Archieves of the Photogrammetry, Remote and Spatial Information. 2022;**5**:15-20. DOI: 10.5194/isprs-archives-XLVI-5-W1-2022-15-2022

[5] Cechavicius R, Visockiene JS, TumelieneInt E. Three-dimensional (3d) modelling of the pedestrian bridge over the river neris. International Archieves of the Photogrammetry, Remote and Spatial Information. 2022;**5**:47-50. DOI: 10.5194/isprs-archives-XLVI-5-W1-2022-47-2022

Section 4

Transport and Infrastructure

Chapter 6

Using Geospatial Analysis to Assist with Clean Vehicle Infrastructure

Yongqin Zhang and Kory Iman

Abstract

Fuel-based vehicles are a significant source of greenhouse gas emissions in America. Electric vehicles (EVs) offer a potential solution to this problem as a cleaner transportation technology. However, to promote the use of EVs, it is crucial to establish a robust infrastructure of charging stations. This research utilized a multi-factor geospatial analysis to address the complex problem of establishing an optimal EV infrastructure. Several factors, such as land use, demographics, and employment centers, were individually analyzed to determine their suitability for EV charging stations. By considering both positive and negative impacts of these factors, we scored them using geospatial analysis to identify the optimal locations for installing charging stations.

Keywords: geospatial analysis, electric vehicles, charging stations, clean vehicles, clean vehicle infrastructure

1. Introduction

Gasoline has been the primary fuel for transportation in the United States over the past 100 years [1]. In 2021, petroleum products accounted for about 90% of the total energy use in the U.S. transportation sector. Long-term use of fossil fuels deteriorated urban air quality, with road transportation being responsible for 70% of pollutants and 40% of greenhouse gas emissions in urban areas [2]. In American, more than half of Americans (166 million) live in counties with unhealthy air quality conditions [3]. Alternative forms of energy such as solar, wind, hydrogen fuel cells, and electricity have been considered as potential energy sources. Electric vehicles (EVs) are powered by electricity which enables them not to introduce harmful pollutants into our atmosphere like gasoline vehicles. EVs convert approximately 59–62% of the electrical energy at the wheels, whereas conventional vehicles only convert about 17–21% of gasoline energy. EVs emit no tailpipe pollutants [4] and have gained support as a strong alternative candidate for future fuel transportation due to the fact of not introducing harmful pollutants to the atmosphere. The U.S. federal government has started some incentive programs to encourage the purchase and use of EVs [5, 6]. Some relevant policies and several incentive programs have been released to ease dependence on gasoline consumption, including purchasing tax credits and installing EV charging stations. These incentives have been adopted by state and local governments. EV charging stations are typically installed by various entities, including governments, companies, and other organizations, to demonstrate their commitment to promoting cleaner transportation options. This support for EV infrastructure is essential to encourage the widespread adoption of electric vehicles and reduce reliance on gasoline-powered transportation.

Although this promising transportation option is available in some places, most of the programs for EV charging stations lack a comprehensive analysis of the locations and infrastructure is not yet in great supply. Many programs install charging stations in urban areas at popular places such as city centers, shopping areas, train stations, and university campuses. More scientific research is needed to better understand where EV charging stations should be located, and provide sound solutions for the establishment of a robust EV charging station infrastructure. In Greater Chicago Area, an agent-based decision support system for electric vehicle charging infrastructure deployment was investigated for the four surrounding counties [7]. The research identified patterns in residential EV ownership and driving activities to facilitate the strategic deployment of new charging infrastructure. An equilibrium-modeling framework was developed to explore interactions between the availability of public charging opportunities, prices of electricity, and destination and route choices of EVs at regional transportation and power transmission networks [8].

The transportation industry has benefited greatly from the use of GIS to help solve complex transportation-related issues and plan the infrastructure of EVs [9–11]. GIS provides a variety of geospatial analysis tools that allow transportation practitioners to create spatial models that can provide answers to challenging transportation questions. It has been used to identify new transportation corridors, determine the socioeconomic and environmental impacts of future transportation facilities, track the construction progress of transportation projects, and many others. GIS analysis also has been used to identify prime locations for EV charging stations. GIS has been used to analyze grid impact of EVs and origin-destination to model spatial and temporal characteristics of EV charging loads [12]. Grid partition has been used in minimizing the distance to the charging station, zoning the planning area, and selecting the best locations for each partition with the considerations of traffic density and charging station capacity constraints [13]. A GIS multi-criteria analysis method was developed to map optimal locations for EV charging stations in Athens, Greek [14]. The method uses a number of different weighted parameters such as population, points of interest, income, and parking distance to map optimal locations. A GIS site suitability model was also used to locate EV charging stations on public facilities in Los Angeles County, CA [15]. Chen et al. [16] have developed a regional methodology to locate EV charging stations through the use of a regression equation that can predict parking demand in urban areas.

These research methods aim to solve the same complex problem of identifying optimal geographic locations for new EV charging stations. Each uses a defined set of weighted demand factors within a spatial model to determine prime locations and proves to be a success in determining locations for EV charging stations. However, these studies used a limited amount of input demand factors to identify optimal locations. The limited use of input demand factors targets specific facilities to install EV charging station areas. The network provides charging stations to a very selective portion of the driving population. These previous studies have only considered a limited number of demand factors. At the same time, the walking distance between an EV charging station and the desired destination tended to be overlooked. Many studies just used an assumption factor for this variable [16]. Identifying optimal locations for EV charging stations is a complex process that involves various factors. These limited approaches may not accurately capture the full range of variables that influence the adoption and usage of electric vehicles. Therefore, it is crucial to conduct more comprehensive analyses that consider a broader range of demand factors to identify the most suitable locations for EV charging stations.

The purpose of this research was to conduct a comprehensive analysis of the various factors that influence the installation of EV charging stations. To achieve this, we developed a multi-factor geospatial method that evaluates a range of positive and negative impacts on potential locations. By considering these factors, our method aims to identify the most suitable locations for EV charging stations to provide widespread availability for the driving public.

2. Multi-factor geospatial method

As a new vehicle system, the current number of public EV charging stations is insufficient to meet the demand, in contrast to the vast number of gas stations available. The average range of EVs per charge is much less than that of conventional gasoline vehicles. This lack of infrastructure presents a challenge for the widespread adoption of electric vehicles, and also leads to range anxiety and limited access to charging stations. Range anxiety (driver's fear that EV has insufficient battery charge to cover the road distance before the destination is reached) prevents this technology from being adopted rapidly by the traveling public. EVs can provide up to 100 miles of distance on a fully charged battery [17]. According to the U.S. Department of Transportation Federal Highway Administration, 100 miles is ample for 90% of trips generated by the traveling public in the United States [18]. Longer trips would require proper planning to ensure the availability of EV charging stations to one's destination. Range anxiety is a common feeling among EV owners that they may be unable to reach their destination before running out of battery power. This phenomenon is enhanced by driving habits, excessive speed, and weather conditions where battery power is used to heat or cool the cabin of the vehicle [19].

In this research, we used a GIS suitability model to evaluate multiple factors and determine the optimal locations for charging stations. This GIS model intakes input demand factors through a series of spatial analysis procedures to generate an output demand factor grid. To run the model and perform the spatial analysis properly, all the input demand factors need to be collected and prepared or spatially interpolated to the right spatial format. Geospatial analysis of these input demand factors are an important step in modeling. A series of data sets that characterize the physical and urban features of the study area are analyzed first to identify trip attractions by the driving public. Highly desirable and most visited locations are usually identified as employment centers, shopping districts, major transportation hubs, public facilities, and recreational areas. Urban and environmental factors can have negative or positive factors to the driving public. A comprehensive analysis of these location factors is performed and a scoring system is developed to evaluate the contributing impact of each factor on the charging station. To prevent any of the input factors from dominating the scoring system, a uniform weighted system is developed to distribute weight to each input demand factor. This systematic scoring method provided the necessary input parameters that were utilized during the analysis to determine suitable and unsuitable areas for EV charging stations.

Input demand factor is evaluated through the use of a GIS suitability model. The suitability model intakes input demand factor through a series of spatial analysis

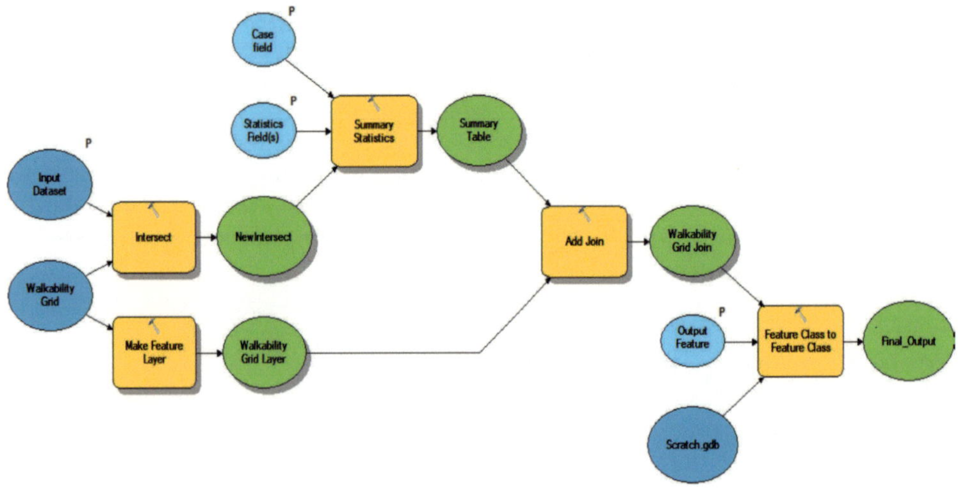

Figure 1.
Input demand factor model.

procedures to generate an output demand factor grid. **Figure 1** illustrates each spatial analysis step used in the GIS suitability model.

The first step is to intersect the input demand factor layer with a walkability grid. A walkability grid indicates how far a commuter would be willing to walk to a particular destination (charging station). A 0.5 × 0.5 mile walkability grid was created in this research as a comfortable walking distance (0.25 mile) for commuters to the charging stations [20].

The second step is to perform a quantile classification to produce a master grid layer. The walkability grid created in step one was then applied to intersect each input demand factor for the study area. The ID of the walkability grid is assigned to associate each individual input demand factor with the walkability grid. A summary statistics is generated using the attribute table of the newly generated input demand factor grid. Based on the summary statistics, a new composite index score is calculated for each grid. This procedure took each individual grid square and tallied up all the input demand factor index scores to produce a composite index score for that particular grid square. For example, if a single grid square contained the following land use types: residential (−4), commercial (+5), and mixed use (+5), the summary statistics would calculate the composite output score to be −4 + 5 + 5 = 6. This composite index score of 6 would then be associated with its grid's id. Each input demand factor grid is added to the intersect tool in ArcMap to produce a master grid layer. This process allowed for each individual index demand factor layer score generated by the model to be combined into one layer for further analysis. A new attribute field called 'factor score' is added to the master grid layer to hold the initial calculated index demand factor score. This calculation is conducted using ArcMap's field calculator which allowed for an equation to be written that summarizes all attribute fields representing each input demand factor layer's composite scores.

The resulting summary statistics table is then joined back to the walkability grid to produce a final output demand factor grid. The final output demand factor grid is then exported out to a new feature class and stored in a file geodatabase. This spatial procedure is then repeated again for each individual input demand factor.

3. Case study

The study area of this research project is in the Utah state, United States, where EV technology has gained public's attention as clean transportation to alleviate the impacts on Utah's air quality. The Salt Lake and Provo/Orem metropolitan areas are part of a unique mountainous region called the Wasatch Front in the State of Utah. The State of the Air reported serious pollution in both short-term and long-term particle air pollution in this area [21].

Salt Lake and Provo/Orem metropolitan areas are listed on the top of short-term air pollution, ranked 6th and 9th, respectively. Atmospheric inversion is the major factor causing air pollution in the area [22]. During winter time, cold air is caped under warm air and traps air pollutants near the valley floor (**Figure 2**). The harmful small particles of air pollutants, such as nitrogen oxides, sulfur dioxide, carbon monoxide, and ozone accumulate in the cold air above the safety levels defined by the U.S. Environmental Protection Agency [23]. The Utah Department of Environmental Quality reported that fuel-based vehicles produced 60% of the polluting particle matter [24]. Exposure to poor air quality conditions has substantial adverse effects on human health, especially for people who have respiratory and cardiovascular conditions. A variety of health issues in Utah is found be to associated with poor air quality. Struggling against air pollution is a longtime task for the state. The Intermountain Air Quality and Health Group was established in 2014 to address the escalating evidence. Intermountain Health Care encouraged its employees to utilize public transportation which contributed to declining emissions of 3.5 million pounds [23]. The State of Utah developed several programs to combat poor air quality along the Wasatch Front, such as the statewide TravelWise program that aims to reduce air pollution by providing alternative transportation ideas. These ideas range from alternative work schedules, active transportation, carpooling, public transit, skip the trip, teleworking, trip chaining, and trip planning [25]. This program encourages employers and citizens to participate in activities that will reduce air pollution throughout the Wasatch Front. There are approximately 106 EV charging stations throughout the State of Utah [18]. Only 77 of those EV charging stations are located along the Wasatch Front where 77%

Figure 2.
Atmospheric inversion in Salt Lake City [22].

of Utah's population resides [26]. To combat range anxiety and increase EV usage, a well-developed robust EV charging station infrastructure needs to be established.

This project took the Wasatch Front in Utah as a case study for its significant role of being the economic hub of the state, being the most populated area in the state, and presenting a large portion of work-related trips from nearby counties [26]. 17.2% of the workforce in Utah County and 47% in Davis County work commute from outside of the county [27]. The commute tends to increase with the growth of the population and economy.

A series of data sets, 34 GIS polygon data layers in total, that characterize the physical and urban features of the study area were collected from various government agencies. The focus was to identify GIS datasets that exhibited a high trip attraction by the driving public. Although this research project focused on locations that matter to the driving public, both positive and negative factors were taken into account. Special attention was given to environmentally sensitive areas along the Wasatch Front to prevent these areas from being identified as prime locations in the analysis. The datasets were compiled and preprocessed to GIS data format.

All data layers were given a weighted score based on their positive or negative influence on an EV charging station. The scoring system for each input demand factor was broken down into the following categories: high suitability (score of 5), moderate suitability (score of 3), low suitability (score of 1), and unsuitable (negative score). These weighted spatial input parameters can be taken into a suitability model to determine the final suitable spatial outcome.

The factor score in the master grid was then multiplied by the physical feature index score (1 for valley floor and − 1 for mountains and waterbodies) which resulted in a series of positive and negative final index demand scores. The resulting negative scores indicated that these grid squares fell among mountainous and waterbody areas that would be unsuitable for an EV charging station. Grid squares that contained positive scores were indicators of suitable valley floor locations for EV charging stations. These positive final index demand scores were further evaluated to determine a proper classification that would assist in identifying prime locations for EV charging stations. An unsupervised classification within ArcMap was initially utilized to first visualize the spatial pattern of the final index demand factor scores. A series of supervised classifications were then applied to the final index demand factor scores and viewed on the top of an aerial photograph of the study area. Visual checks were applied to areas known to have a high attraction for EV charging stations. The resulting scoring patterns in these areas were used to establish a uniform supervised classification system that was used to display high, moderate, and low suitability areas for EV charging stations throughout the study area.

4. Geospatial scoring and mapping processes

Table 1 details the scoring for each input demand factor. The input demand factors are classified into four major categories: environmental factors, demographics, infrastructure, and land use. The environmental factors describe the physical features of the landscape in the study area. Demographic analysis was given to have a perspective on the population, employment status, and trip destinations. Infrastructure entails the locations that the driving public may visit. The land surface is classified based on its purpose of usage.

The study area has several landscape types including mountains, lakes, rivers, and wetlands. To delineate the landscape types for the scoring process, a digital elevation

Category	Factor	Score	Score definition
Environmental			
Physical features	Mountains	−1	Unsuitable
	Water bodies	−1	Unsuitable
	Valley floor	1	Suitable
	Major rivers	−1	Unsuitable
	Wetlands	−5	Unsuitable
Demographics			
Population (added factor)	High	2	High suitability
	Low	1	Low suitability
Employment	High	5	High suitability
	Medium	3	Moderate suitability
	Low	1	Low suitability
Trip destinations	High	5	High suitability
	Medium	3	Moderate suitability
	Low	1	Low suitability
Infrastructure			
	Airports	5	High suitability
	Existing EV charging stations	−5	Unsuitable
	Gas stations	5	High suitability
	Golf courses	5	High suitability
	Government fuel sites	3	Moderate suitability
	Government offices	5	High suitability
	Health care facilities	5	High suitability
	Libraries	5	High suitability
	Major attractions	5	High suitability
	Major intersections	5	High suitability
	Major roads	5	High suitability
	Major parking garages	5	High suitability
	Park & ride lots	5	High suitability
	Parks	5	High suitability
	Places of worship	5	High suitability
	Post offices	5	High suitability
	Schools	5	High suitability
	Senior centers	5	High suitability
	Shopping malls	5	High suitability
	Ski resorts	5	High suitability
Land use			

Category	Factor	Score	Score definition
	Agriculture	−5	Unsuitable
	Open space	−5	Unsuitable
	Commercial	5	High suitability
	Mixed use	5	High suitability
	Industrial	5	High suitability
	Government/ institution	5	High suitability
	Residential	−4	Unsuitable
	Sensitive areas	−5	Unsuitable

Table 1.
Input demand factor scoring.

model was first applied to separate the landscape into two categories: mountains and valley floor. The subsequent layer was converted to a polygon and intersected with a major lake feature class to get polygons of mountain and lake areas. A score of −1 (unsuitable) was assigned to mountain and lake areas, a score of 1 (suitable) was assigned to valley floors, and a score of −1 (unsuitable) was assigned to a separate polygon layer that contains major rivers. A score of −5 (unsuitable) was given to protected wetlands in the study area, which protects these locations from being used for potential EV charging stations.

A score of 5 (high suitability), meaning a strong attraction to the driving public, was assigned to the majority of the input demand factors for infrastructure. Facilities that are often visited by the public belong to this category and are assigned a score of 5. These infrastructures are identified as gas stations, airports, health care facilities, libraries, golf courses, government buildings, postal offices, schools, shopping malls, places of worship, senior centers, popular parking garages, park and ride lots, major attractions such as parks and ski resorts, as well as major road intersections and roads. A score of 3 (moderate suitability) was given to government fuel sites since these locations might be unavailable to the traveling public. Adding these sites to the analysis would encourage government agencies that utilize fleet vehicles to invest in EV technology. Existing EV charging stations were the only negative infrastructure input factor that was assigned a −5 (unsuitable) score. The reason for this negative input was to reduce the chance of highlighting areas that already had an EV charging station established.

Analysis of the distribution of demographics assisted with the perspective of socioeconomics in the area. The current and future population trends, employment and employers, and trip destinations were all evaluated for the study area to assign the scoring number for the input demand factors. A quantile classification was run in ArcMap to score the 2040 employment estimates contained in the Traffic Analysis Zones (TAZ). Areas of high employment (score of 5), medium employment (score of 3), and low employment (score of 1) were labeled respectively based on the classification. The 2040 TAZ population estimates were analyzed using a different approach. Most EV owners will generally have a home EV charging station installed in their homes for convenience. This factor needed to be taken into consideration since the targeted locations for this research project focused on trips away from home. To avoid highlighting residential areas, the population was analyzed on a regional scale and used as an added factor to the overall analysis. This allowed for

Company	Industry	Employment	Location
Intermountain Health Care	Health Care	20,000+	Wasatch Front
State of Utah	State Government	20,000+	Wasatch Front
University of Utah	Higher Education	20,000+	Wasatch Front
Brigham Young University	Higher Education	15,000–19,999	Wasatch Front
Walmart Associates	Warehouse	15,000–19,999	Wasatch Front
Hill Air Force Base	Federal Government	10,000–14,999	Wasatch Front
Davis County School District	Public Education	7000–9999	Wasatch Front
Granite School District	Public Education	7000–9999	Wasatch Front
Utah State University	Higher Education	7000–9999	Cache County
Smith's Food & Drug Centers	Grocery Stores	7000–9999	Wasatch Front

Table 2.
Largest employers in the state of Utah.

high to moderately populated areas to receive additional points to better identify potential locations of high EV charging usage. Areas of high to moderate population were given a score of 2, while areas of the low population received a score of 1. The Utah Department of Workforce Services identifies Utah's largest employers on an annual basis [28]. Out of the top 10 employers in Utah, 9 of them reside along the Wasatch Front as seen in **Table 2**. These employers generate a high number of work-related trips and therefore were given a score of 5 (high suitability).

The State of Utah conducted a Household Travel Survey in 2012 to better understand the travel patterns of the driving public. The trip destination portion of this study was utilized in this research project to identify areas where the traveling public visit the most frequently. A quantile classification was performed to identify areas of trip frequency. The above-average trips are scored 5, a moderate amount of trips are scored 3, and low trips are scored 1.

Land use plays a direct role in determining the types of destinations that are taken by the driving public, therefore the attractiveness of the land use type determines the scores for this input factor. Plans for general land use of each county was referenced to project current and future land uses. Land use types that offer services and employment usually attract a high number of destination trips, such as lands used for commercial, industrial, government, institutional purposes, or mixed use of these functions were scored 5 with high suitability for charging stations. On the other hand, environmental land use types such as agriculture, open space, and sensitive areas are unsuitable for installing charging stations and were scored −5. Charging stations for residential use are generally installed in private residences for owners of vehicles. Residential areas were unsuitable and therefore scored −4. This scoring process eliminates the possibility of residential areas being highlighted but also prevented adjacent optimal land use types from being overshadowed.

Each input demand factor was produced as a separate grid to illustrate destination attractions and spatial distribution of the demand factor within the study area. Each of the resulting input demand factor grids was composed of 16,790 individual 0.5 × 0.5-mile grid squares. **Figure 3** illustrates all the 28 input demand factor grids. In these maps created, input factors with positive score are shown in blue whereas negative scores are colored orange. There were a total of 23 out of 28 input demand

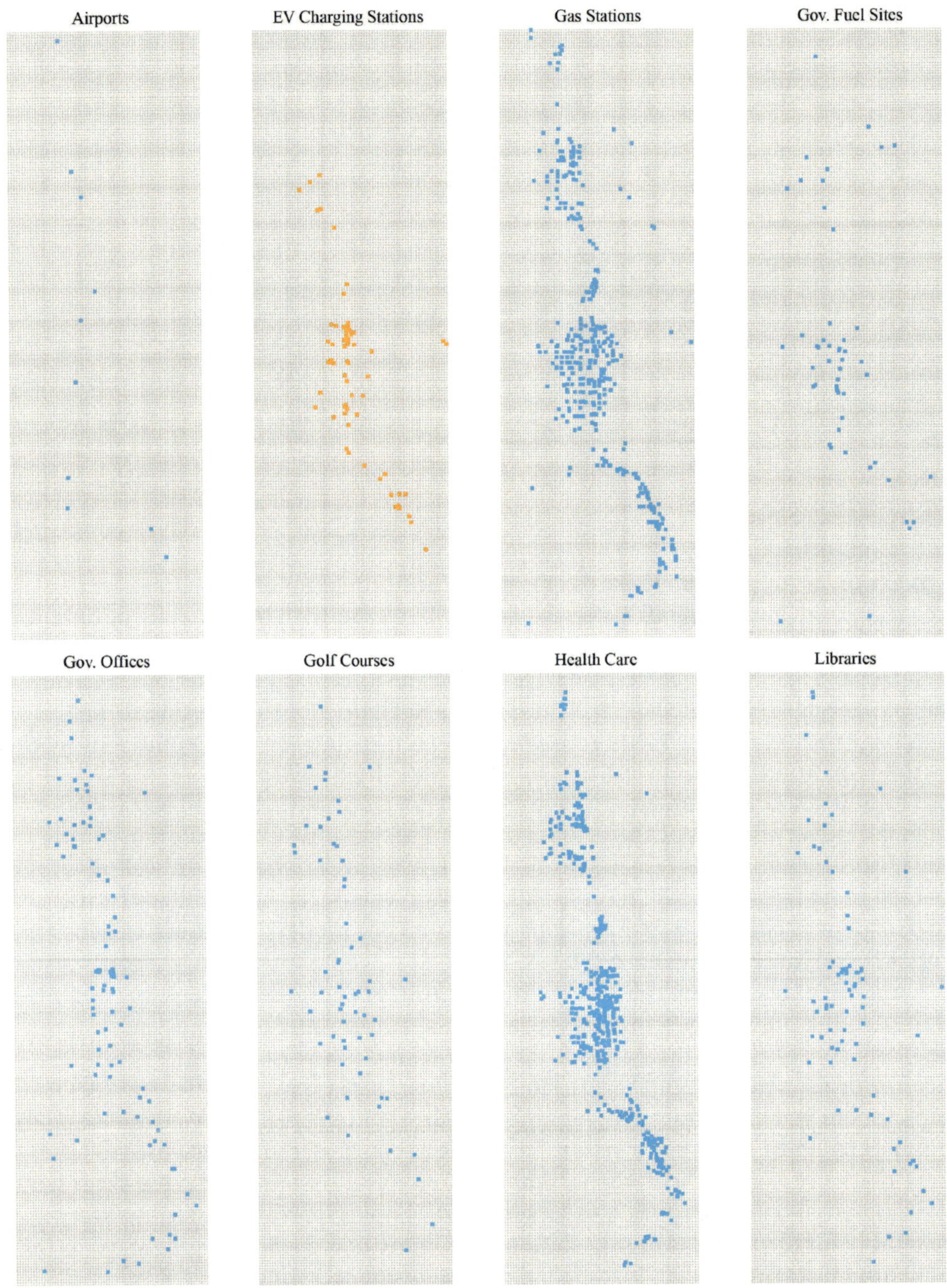

Figure 3.
Maps of the demand factor grids.

factor grids that received a positive score. As shown in **Figure 3**, these positive input demand factor grids are characterized as major transportation hubs, employment centers, shopping districts, public facilities, health care centers, recreational areas, future populations, and trip destinations. The resulting positive input demand factor grids played an important role in determining the final output of this suitability analysis. Only 3 input demand factor grids received all negative scores which included existing EV charging stations, rivers, and wetlands. These particular negative input demand factor grids helped to deter these locations from being identified in the final analysis. Also, 2 of the input demand factor grids contained both positive and negative scores. These specific input demand factor grids represented land use and physical features. Additionally, these input demand factor grids assisted in identifying land area types that were suitable and unsuitable for locating EV charging stations within the study area. The scoring system contained in these input demand factor grids became the vital stepping stone that allowed for the final results to be generated. Converting the initial input datasets into these uniform input demand factor grids allowed for a final composite score to be spatially calculated. The final composite scores were used to display the final results as a hot spot map.

A comparison of each input demand factor grid revealed a unique spatial distribution pattern. This unique pattern outlined where these input demand factors were concentrated along the Wasatch Front area. The majority of these input demand factors were located in urbanized centers or adjacent to major transportation facilities. This spatial pattern became very apparent when these input demand factor grids were compared with Utah's major transportation facilities. I – 15 is the principal interstate that runs north and south through the entire State of Utah. Many of the input demand factors throughout the Wasatch Front were generally concentrated near this major transportation facility. This spatial concentration allows for greater access to these particular input demand factors. Prime examples that illustrate this spatial phenomenon can be seen in the following input demand factor grids: government offices, gas stations, libraries, major attractions, major employers, park & ride lots, post offices, and senior centers. Salt Lake County differs a bit from other counties along the Wasatch Front when it comes to core concentrations of input demand factors along I – 15. This distribution difference is due to a higher population density and the effects of urban sprawl that has led to a greater spatial expansion of input demand factors in this area. Salt Lake County is home to more than one interstate; I-15, I-80, and I-215, which provides greater accessibility than any other surrounding counties in the study area. This wider spatial distribution of input factors within Salt Lake County is also apparent in many of the resulting input demand factor grids. Higher concentrations of input demand factors will allow for a greater probability of locating a greater number of EV charging stations within Salt Lake County. Other input demand factors that did not necessarily follow the core concentration concept were factors that provided services to smaller geographic areas such as local parks, places of worship, health care facilities, schools, and ski resorts. These input demand factors had a broader geographical stretch compared to other input demand factor grids. This allows for areas outside the core concentration area to be considered and assists in developing a more expansive EV charging station system. These individual input demand factor grids help portray each input factor's influence in the overall analysis.

5. Hot spot mapping

A hot spot map was produced to demonstrate locations with high, moderate, and low potentials for installing an EV charging station for the study area (**Figure 4**). High suitable areas were highlighted in red on the map. The areas have an index demand score of 50 or greater, and contain the highest mixture of input demand factors. These are driving public frequently visited areas and are ideal for installing EV charging stations. Downtown business districts, shopping districts, employment hubs, public facilities, health care centers, park and ride lots, parking garages, and major roads and intersections, attractions, recreational areas etc., are all hot spots for EV charging stations. The yellow areas represented moderate suitable areas. This class displayed an uneven mixture of input demand factors. Moderate demand areas are secondary choices for charging station when prime areas cannot be located in surrounding areas.

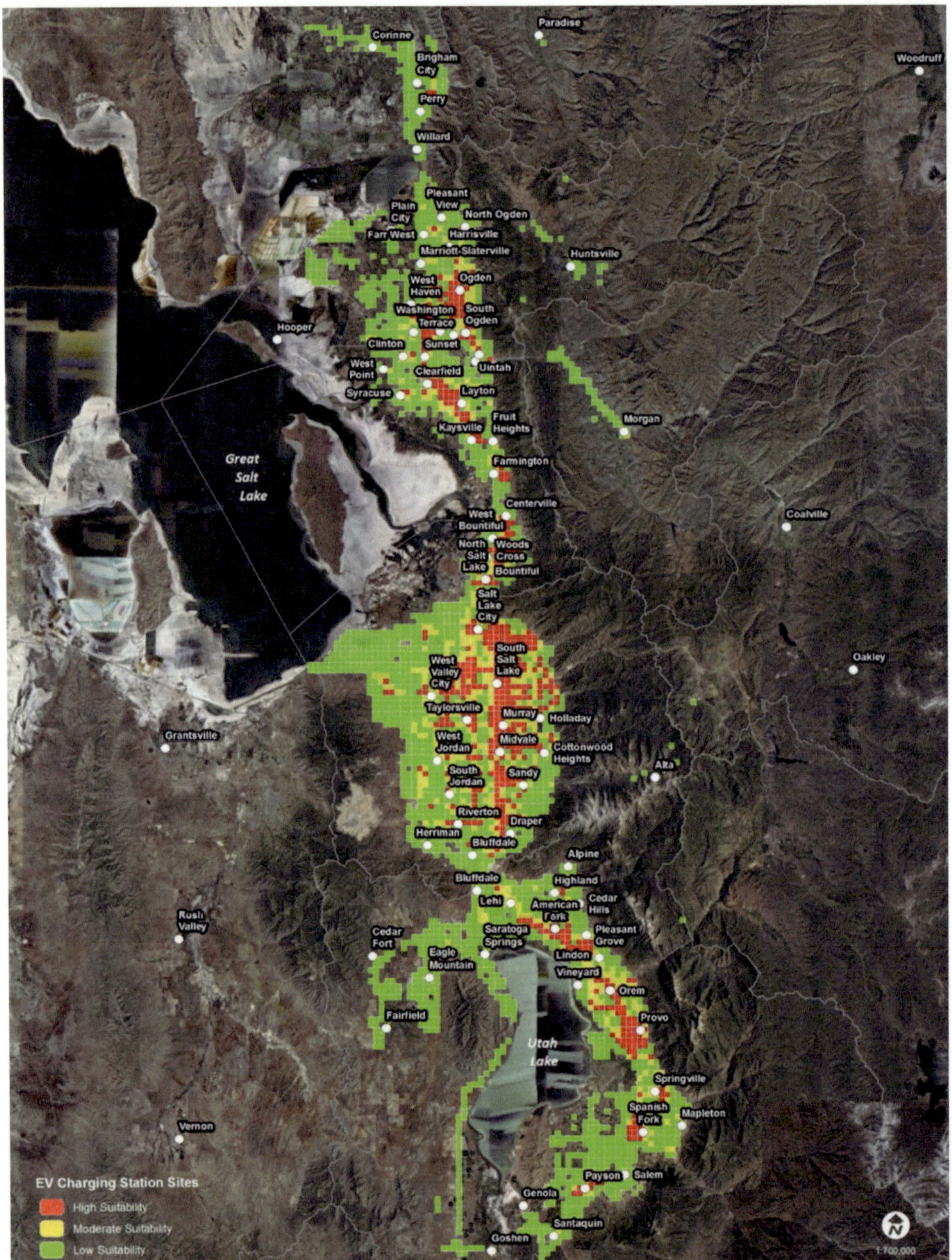

Figure 4.
Optimal locations to install EV charging stations.

Influence	Input demand factor	Percentage of
High	Major parking garages	99
	Shopping malls	99
	Senior centers	91
	EV charging stations	90
	Schools	88
	Gas stations	88
	Libraries	86
	Health care facilities	83
Moderate	Post offices	79
	Major attractions	78
	Government offices	76
	Park and ride lots	74
	Government fuel sites	70
	Places of worship	53
	Parks	52
	Major road intersections	50
	Major roads	50
Low	Airport	
	Ski resorts	
	Golf courses	

Table 3.
Influence of the demand factors.

Examples are small business parks near larger residential neighborhoods, and areas that are seeing population and employment growth. Low suitability areas were colored green, where showed a low mixture of influential factors and surrounded by low suitability factors such as residential neighborhoods, open space, and agricultural areas.

The multi-factor modeling system developed in this research classifies the hierarchical importance of each input demand factor, and identifies suitability of locations for installing EV charging stations. The ranking for each input demand factor was tested against the total number of input grids. The testing showed that the majority of the input demand factors had an average influential percentage of 80%, as listed in **Table 3**, indicating the model's ability to produce reliable results [29].

6. Conclusion

This research was to develop a multi-factor methodology for identifying optimal locations for installing EV charging stations. A comprehensive geospatial analysis was conducted on multiple input demand factors to develop a scoring system and locate areas that are most frequently used by the driving public. This scoring system was able to calculate a composite score for each location and spatially classify influential factors to identify suitable locations for EV charging stations. The locations with high, moderate, and low suitability for installing EV charging stations was identified and spatially delineated. The hot spot map of suitable locations can assist decision-makers with developing a strong EV charging station infrastructure. Geospatial analysis plays a vital role in the comprehensive evaluation and determination of the scoring for each input demand factor and thereby identifying optimal locations for installing charging stations.

Author details

Yongqin Zhang* and Kory Iman
Delta State University, Cleveland, MS, USA

*Address all correspondence to: yzhang@deltastate.edu

References

[1] Xu K, Yi P, Kandukuri Y. Location selection of charging stations for battery electric vehicles in an urban area. International Journal of Engineering Research and Science & Technology. 2013;**2**:15-23

[2] EEA. The Contribution of Transport to Air Quality - TERM 2012: Transport Indicators Tracking Progress towards Environmental Targets in Europe. Luxembourg: European Environment Agency; 2012

[3] American Lung Association. 2016 State of the Air. 2016. Available from: http://www.lung.org/assets/documents/healthy-air/state-of-the-air/sota-2016-full.pdf

[4] Plug-in Electric Vehicle Resource Center. Battery electric vehicles. 2015. Available from: https://driveclean.ca.gov/battery-electric [Accessed: March 20, 2022]

[5] Fuel Economy. Federal Tax Credits for Electric Vehicles Purchased in or after 2010. 2015. Available from: http://www.fueleconomy.gov/feg/taxevb.shtml [Accessed: October 28, 2016]

[6] Peterson SB, Michalek JJ. Cost-effectiveness of plug-in hybrid electric vehicle battery capacity and charging infrastructure investment for reducing US gasoline consumption. Energy Policy. 2013;**52**:429-438

[7] Sweda T, Klabjan D. An agent-based decision support system for electric vehicle charging infrastructure deployment. In: Vehicle Power and Propulsion Conference (VPPC). Chicago, IL, USA: IEEE; 2011. pp. 1-5

[8] He F, Di W, Yin Y, Guan Y. Optimal deployment of public charging stations for plug-in hybrid electric vehicles. Transportation Research Part B: Methodological. 2013;**47**:87-101

[9] Bailey J, Miele A, Axsen J. Is awareness of public charging associated with consumer interest in plug-in electric vehicles? Transportation Research Part D: Transport and Environment. 2015;**36**:1-9

[10] Namdeo A, Tiwary A, Dziurla R. Spatial planning of public charging points using multi-dimensional analysis of early adopters of electric vehicles for a city region. Technological Forecasting and Social Change. 2014;**89**:188-200

[11] Marco NY, Ghamami M. A corridor-centric approach to planning electric vehicle charging infrastructure. Transportation Research Part B: Methodological. 2013;**57**:172-190

[12] Mu Y, Jianzhong W, Jenkins N, Jia H, Wang C. A spatial-temporal model for grid impact analysis of plug-in electric vehicles. Applied Energy. 2014;**114**:456-465

[13] Ge S, Feng L, Liu H. The planning of electric vehicle charging station based on Grid partition method. In: International Conference on Electrical and Control Engineering, Yichang, China. 2011. pp. 2726-2730

[14] Dimitrios E, Constantinos A, Yannis T, Evangelos M. Spatial exploration of effective electric vehicle infrastructure location. Procedia – Social and Behavioral Sciences. 2012;**48**:765-774

[15] Jin J. Installing Public Electric Vehicle Charging Stations: A Site Suitability Analysis in Los Angeles County, California. Los Angeles, CA: University of Southern California; 2016

[16] Chen TD, Kockelman KM, Khan M. The electric vehicle Charging Station location problem: A parking-based assignment method for Seattle. Transportation Research Record: Journal of the Transportation Research Board. 2013;**2385**:28-36

[17] Alternative Fuels Data Center. All - Electric vehicles. 2016c. Available from: http://www.afdc.energy.gov/vehicles/electric_basics_ev.html

[18] Alternative Fuels Data Center. Developing infrastructure to charge plug-in electric vehicles. 2016a. Available from: http://www.afdc.energy.gov/fuels/electricity_infrastructure.html

[19] Alternative Fuels Data Center. Electric vehicle charging station locations. 2016b. Available from: http://www.afdc.energy.gov/fuels/electricity_locations.html

[20] Yang Y, Diez-Roux AV. Walking distance by trip purpose and population subgroups. American Journal of Preventive Medicine. 2012;**43**(1):11-19

[21] 2016 State of the Air. American Lung Association. 2016. . Available from: http://www.lung.org/assets/documents/healthy-air/state-of-the-ai r/sota-2016-full.pdf

[22] Curtis R. A history of the inversion: A foe that grows stronger. 2014. Available from: http://utahpoliticalcapitol.com/2014/01/22/a-history-of-the-inversion-a-foe-that-grows-stronger/

[23] Joy EA, Horne BD, Bergstrom S. Addressing air quality and health as a strategy to combat climate change. Annals of Internal Medicine. 2016;**164**:626-627

[24] Breathe Utah. Winter pollution: Where does PM2.5 come from? 2015. Available from: http://www.breatheutah.org/winter _pollution [Accessed: October 10, 2016]

[25] State of Utah. Travel wise: Work a schedule that fits your lifestyle. 2008. Available from: https://travelwise.utah.gov [Accessed: April 20, 2022]

[26] Mountainland Association of Governments. Population and demographic data. 2016. Available from: http://mountainland.org/data [Accessed: October 1, 2016]

[27] Mountainland Association of Governments. TransPlan40 - Regional Transportation Plan - 2015 – 2040 Plan for the Provo/Orem Metropolitan Area. 2015. Available from: http://mountainland.org/img/transportation/TransPlan40/TransPlan40.pdf

[28] Utah Department of Workforce Services. Utah's Largest Employers: Annual Average Employment 2014.2015. Available from: https://jobs.utah.gov/wi/pubs/em/annual/current/lrgstemp.pdf

[29] Zhang Y, Iman K. A multi-factor GIS method to identify optimal geographic locations for electric vehicle (EV) charging stations. The International Journal of Cartography. 2018;**1**:1-6

Section 5

Mining Exploration and Monitoring

Chapter 7

GIS and Database Management for Mining Exploration

Jonas Didero Takodjou Wambo and Wenzi Tanni Quinter

Abstract

The mining sector has several tools which permits mining exploration works to be done in the most efficient and time conscious way as possible. This chapter is focus on Geographic Information Systems (GIS) and Database and aims to show how this combined approach can help in mineral exploration. It is all about building a database and GIS system capable first of all to assess if a geological/geographical data taken from the field is valid or wrong and to effectively manage mining data in a GIS designed to positively influence any mining exploration project from its earliest stages to the end of the project. We focus on Meiganga area where we have been involved.

Keywords: GIS, database, spatial analysis, mining exploration, database management

1. Introduction

GIS and database are conceptualized frameworks that are designed to capture, gather, input, store, manage and analyze geographical data. Rooted in the science of geography, GIS integrates many types of data. It analyzes layers of information into visualizations using maps and 3D scenes with this unique capability. GIS reveals deeper insight into data, such as patterns, relationships, and situations helping users make smarter decisions. GIS is specialized in geographical related data unlike database which is able to handle non-geographical data. The application of GIS and Database management in the initial stages of ore mineral exploration has a significant impact on reducing exploration and exploitation prices [1–3]. The effective management of natural resources in many areas of applied geoscience is influenced by obtaining detailed geological/geographical information from various sources [1, 2]. Exploration geologists are generally confronted with various problems mainly related to the difficulty of covering large areas (time and limited access, natural and human risks) leading to a disparity in the information collected and methods of extrapolation and interpolation used for map model [1, 2]. The quality of data collected, is very crucial in determining the success or failure of any mining project activity. Proper handling of this data enters into one of the factors to be respected to ensure their reliability. With the rapid advancements in technology, the management of complex data is somehow easy. Essentially the mining sector has several tools which permits work to be done in the most efficient and time conscious way as possible. GIS coupled with Database management stand out as indispensable way to achieve this goal.

 IntechOpen

The spatial relationship of deposits can be assessed using spatial analysis with the support of a GIS, which can help unravel the geological processes that have been crucial in the formation of a particular type of mineral deposits [4, 5]. Thus, the systematic analysis of spatial data can facilitate the identification of mineral deposit properties within a metallogenic province [4, 6, 7]. Studying the spatial distribution of mineral deposits and the relationship between mineral deposits and determining geological factors using spatial analysis with GIS seems to be a suitable approach that can provide invaluable information at the start of any mining project.

Principally, the technique approach that we will dwell on in the course of this chapter is GIS and Database. The objective of this work is to build a database and a GIS system capable first of all to assess if a geological/geographical data taken from the field is valid or wrong and to effectively manage mining data in a GIS designed to positively influence any mining exploration project from its earliest stages to the end of the project. We focus on Meiganga area where we have been involved.

2. Generalities in GIS, database management

2.1 GIS

A GIS is a system capable of entering, storing, modifying, recording, query, reorganizing, analysis, and display as well as presenting alphanumerically and graphically data located in space to solve a problem, support a decision, and help to plan. It comprises of functional modules allowing to create and modify, to interrogate and to represent cartographically data located in space. There are several components when working on a GIS Project. These consists of setting up a GIS unit, the design, development and information of a reliable database, the implementation of a knowhow for the launch of a dynamic GIS, the use of simple GIS software to fill in the heritage database, acquisition of a suitable hardware platform. Also, there are several basic questions that a GIS should be able to answer such as:

- **Where?** Where is this object, this phenomenon? Where are all the objects of the same type? *(This question allows highlighting the spatial distribution of an object).*

- **What?** What can we find there? *(It is a question of highlighting all the objects or phenomena present on a given territory).*

- **How?** What are the relationships that exist or not between objects and phenomena? *(This is a problem of spatial analysis).*

- **When?** When did the changes occur? What is the age and evolution of a particular object or phenomenon? *(A problem of temporal analysis).*

There are two fundamentally different types of GIS data used in GIS projects:

- **Vector data:** Which are structured data used to store spatial data like lines, points, and polygons. They are based on vectors as opposed to space occupancy raster structures.

- **Raster Data:** A map in a raster data divides the entire study area into a regular grid of cells in a specific sequence. The data completely fill the space of the map.

The power of a GIS comes from the ability to relate different information in a spatial context and to reach a conclusion about this relationship. Most of the information on the world contains a location reference, placing that information at some point on the globe. A GIS, therefore, can reveal important new information that leads to better decision making.

2.2 Importance of GIS in mining exploration

Several authors have studied on the importance of GIS in mineral exploration [3, 8–12]. During mineral exploration vast amounts of spatial data are collected. Geo Science information which commonly consist of geological, geographical, geophysical and geochemical information, are gathered routinely from satellite and airborne sensors in a digital form. Large sets of ground survey data are also stored in digital form as argued by [13, 14]. Accordingly, integration of field survey data, maps, and other information for the purpose of mineral exploration and resource estimation is a very time consuming task. However, GIS can accomplish such a task in a time efficient and cost-effective manner (**Figure 1**). Moreover, the state and federal agencies involved in the mine permitting process are adopting the GIS format as the standard for communicating spatial data.

Although GIS is able to manage data like database management systems, mining exploration projects use to store data in other databases. That is because of the semantic form of data. Meaning that usually, they contain information describes as "*good for spatial purpose*" and other that are used for assessment, management, and validation. Do not forget that, mining companies also use these two types of database for backup of their data and for the distribution.

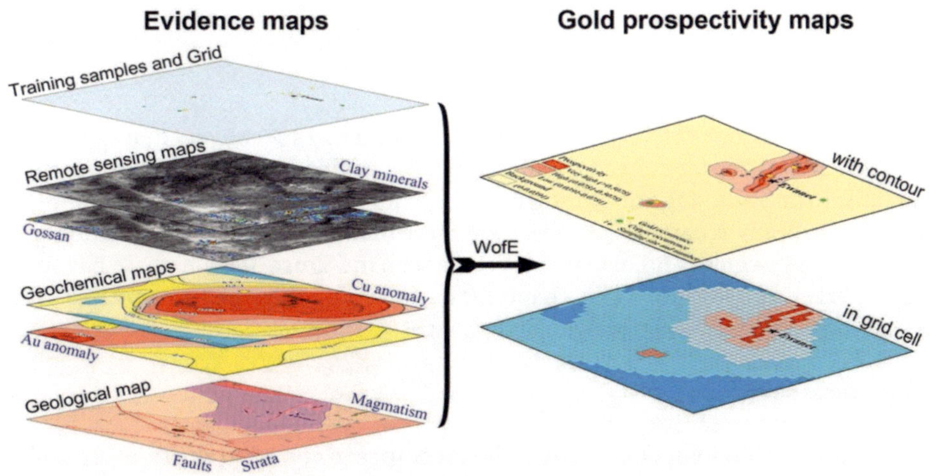

Figure 1.
Synthetic representation of the application of GIS in mineral exploration [15].

2.3 Database management

A database is a collection of related data stored in a manner that enables information to be retrieved as needed. This typically consists of tables (a collection of related records), fields and records. Field is a single category of data to be stored in a database like name and their abbreviations, telephone number, etc. Fields are collection of related fields information in a database. Excel and Access are forms of database software. Databases are important because of their: low level of redundancy, faster response time, lower storage requirements, easier to secure, increased data accuracy. Many types of database exist depending of the level of usage. We will name here the two most widely used in the level of single users and small projects:

A Single-User Database System is located on a single computer and is designed to be accessed by one user. This type of database is widely used for personal applications and very small business.

A Multiuser Database System is designed to be accessed by multiple users (most business databases today).

A database with only one table is sometimes called a flat-file database.

2.4 Importance of database in mining projects

Database are used in mining exploration project because of: (1) Their property to be imported into a GIS or to mining software, mostly in case they contain geographical information system like coordinates; (2) They permit to write and adopt their Standard operations procedures (SOP) by codification which leads the whole of their field work.

Thus, for a sample to enter the database, a name must be given, an identity, coordinates, location, prospect and project name (**Figure 2A**). If a name does not already belong to the database built, an error occurs. The notebook of the geologist is considered as database in some projects. In others, printed spread sheets of excel are used for (**Figure 2B**). So that the physical database exists and the numeric database exists also.

3. Methodology

3.1 Main phases

Given that GIS and database are meant to respond to a specific need, it is imperative that, objectives of the task ahead should be clearly outlined to suit expectations most accurately. In an early stage of mining projects wherein stream sediment sampling and large scale mapping is carried out, there are three main phases which are Planning, Field work, Assessment and data validation as seen in **Figure 3**. Each phase has special tools, expertise and logistics. This chapter is design going through those phases of the work laying emphasis on the Meiganga toposheet.

3.2 Technical settings

Every project has its specifications when it comes to choosing the types of tools and software in order to achieve the work. In this case the equipment used include: the GIS and database equipment which consist of a laptop, notebooks, and GPS. GIS software like ArcGIS 10.5 can be chosen to build maps and make geospatial analysis. Basecamp is also used with the aim to send maps and planned points inside the GPS.

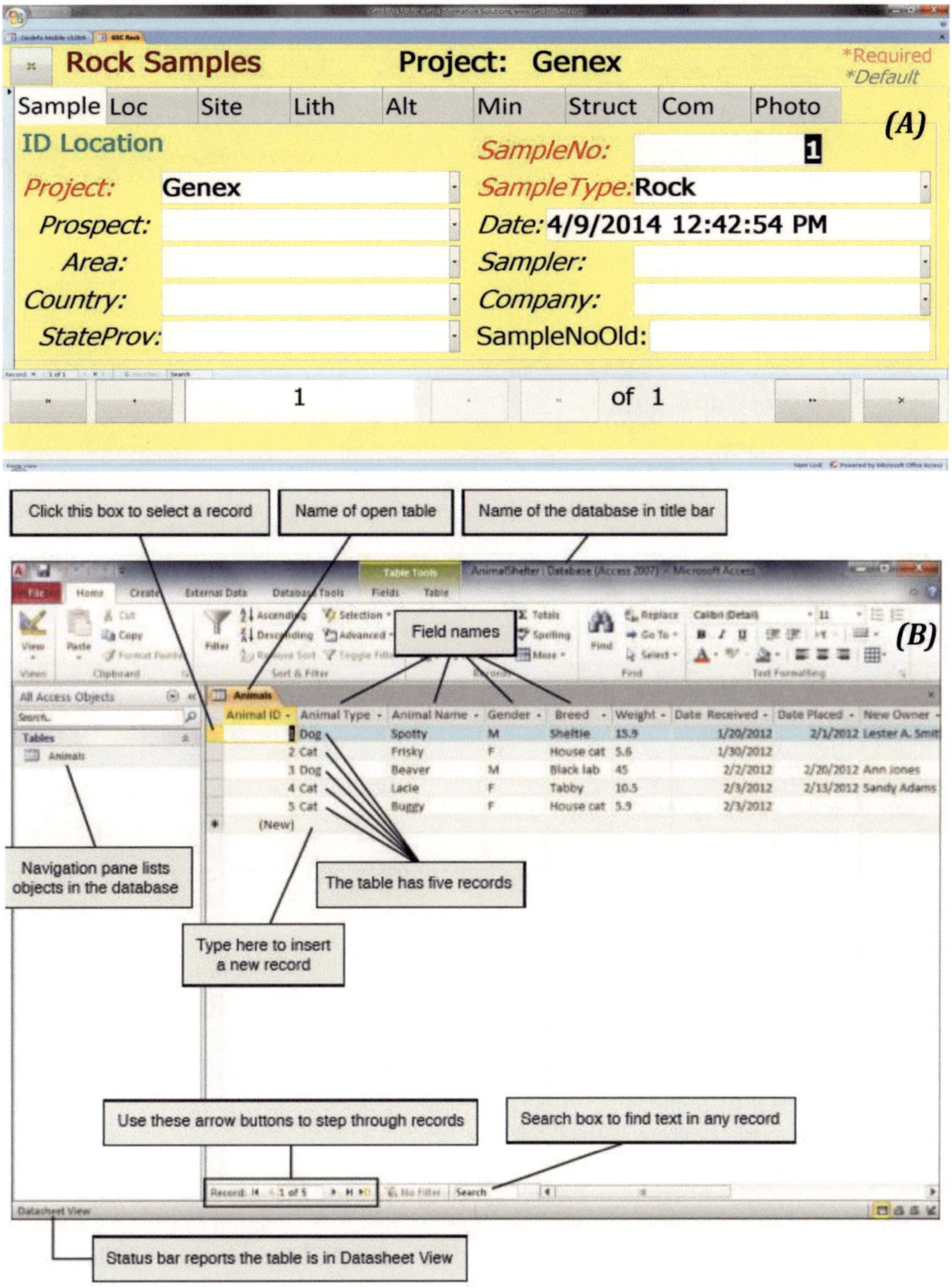

Figure 2.
Database formats: (A) presentation of geographic information and SOP and (B) presentation of non-geographical data.

The same software can be used to extract prospectors' tracking and recorded points, the coordinate system used in the building of maps can be WGS 1984, with UTM as units. As regarding database, Microsoft Access can be used to build tables and record related field information. The software is also used to record all the codes employed during the work i.e. rock codes, initials of the geologists, abbreviations of the targeted

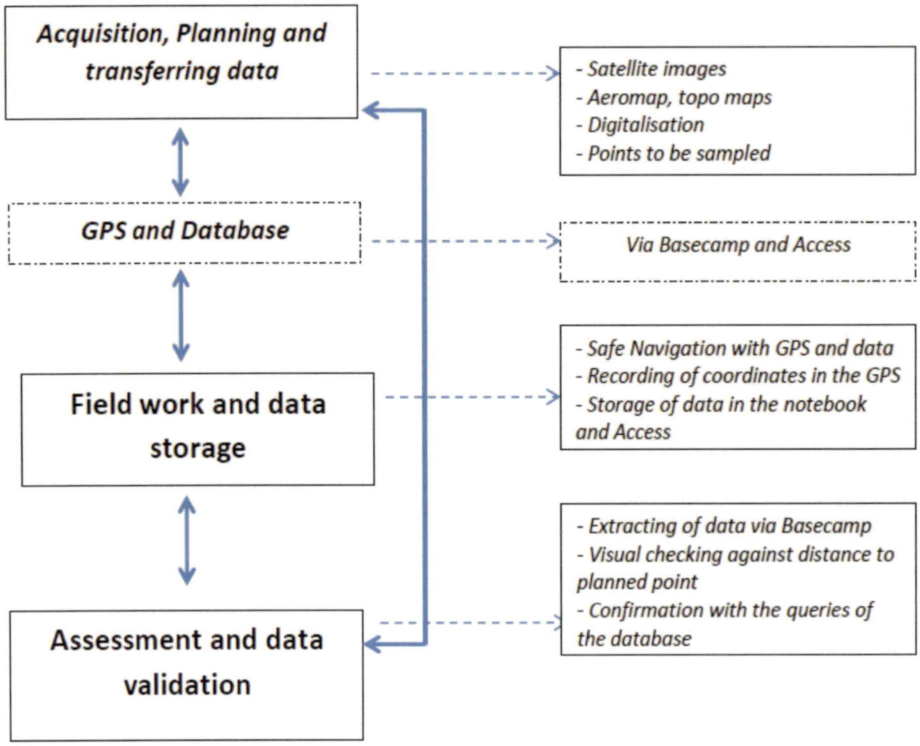

Figure 3.
Flowchart of the methodology of study.

zone, and other information. For airborne geophysical data, Geosoft can be used for analysis.

3.2.1 Acquisition, planning/transferring

The planning phase entails collection of data that will be used by the project. It ends with the transfer of data into a GPS and into a PC. Data collected can contain geographical and non-geographical features.

3.2.2 GIS and maps

At the end of the planning phase, three types of resources related to the GIS are available i.e. physical maps, a raster map inside a GPS, and the GIS inside the PC. The GIS of the project should contain all the necessary maps for the team to work on the field. It consists of the information concerning basemaps, topographic maps, Digitital Terrain Model (DTM) or hill shade, hydrography, administrative units and planned points. ArcGIS software can easily store the GIS. Here some fields and maps can be activated or deactivated, in order to find additional information for the field team. The GIS is used to convert all the above maps into one raster in order to be manipulated in other systems. When converted, the maps are printed to obtain physical maps (**Figure 4**). The same maps can be sent into GPS for the navigation through Basecamp software.

The term basemap is seen often in GIS and refers to a collection of GIS data and or orthorectified imagery that form the background setting for a map. Satellite images can

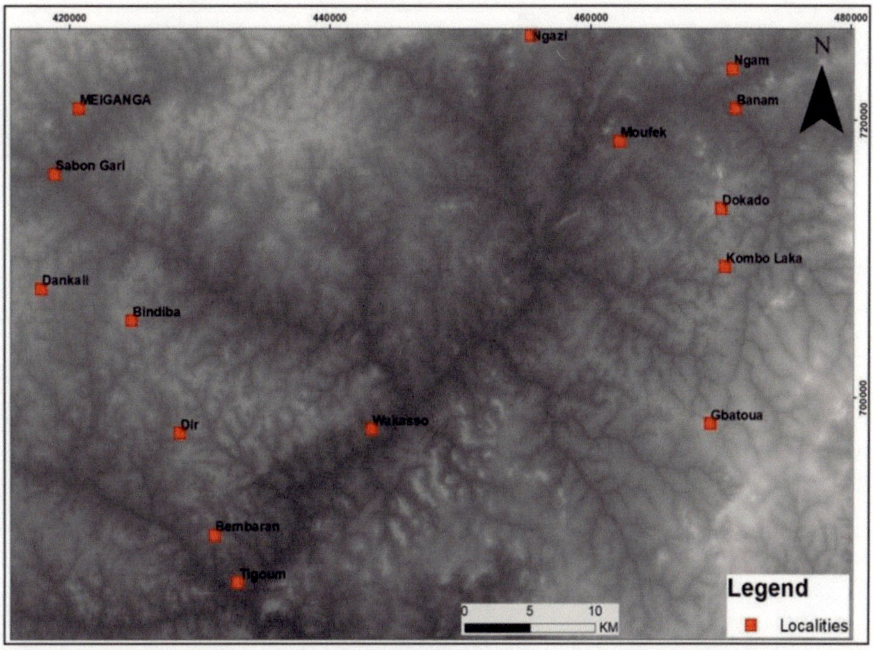

Figure 4.
Satellite image (SRTM) of the study area with 30 m of resolution.

be collected from free verified available platforms like United States Geological Survey (USGS). They store geographical information, localization, and many others important features. The function of the basemap is to provide background detail necessary to orient the location of a map. Basemaps serve as a reference map on which you overlay data from layers and for geologists to be able visualizes geographic information.

The shaded relief is also used to highlight the forms of the relief and enhance the quality of the map (**Figure 5**). The hillshade function produces a gray scale 3D representation of the terrain surface, with the sun's relative position taken into account for shading the image. It is a technique for visualizing terrain determined by a light source and the slope and aspect of the elevation surface; this tool creates a shaded relief raster from a raster. The illumination source is considered to be at infinity. The hillshade raster has an integer value range of 0–255 m.

Figure 6 shows hydrography map of the study area. It is a survey map which reveals the waterway system of the study area. The hydrographic pattern is dense with the main stream being the Lom River with the primary tributary being; mikila and several second order streams.

A topographic map is a detailed and accurate illustration of man-made and natural features on the ground such as roads, railways, power transmission lines, taken from the national institutions are put on the basemaps and used for geo-referencing and digitalization of roads and streams. Topographic maps are raster maps like basemaps. They store information related for topography information. Localization and streams are also a part of them (**Figure 7**). This help for digitalization.

Digitalized data are built by the GIS unit. The team digitalizes streams and roads mainly based on the basemap and the topographic map. The result here is a vector map. The objective is to have a digitalized feature as they are easy to manage than raster maps.

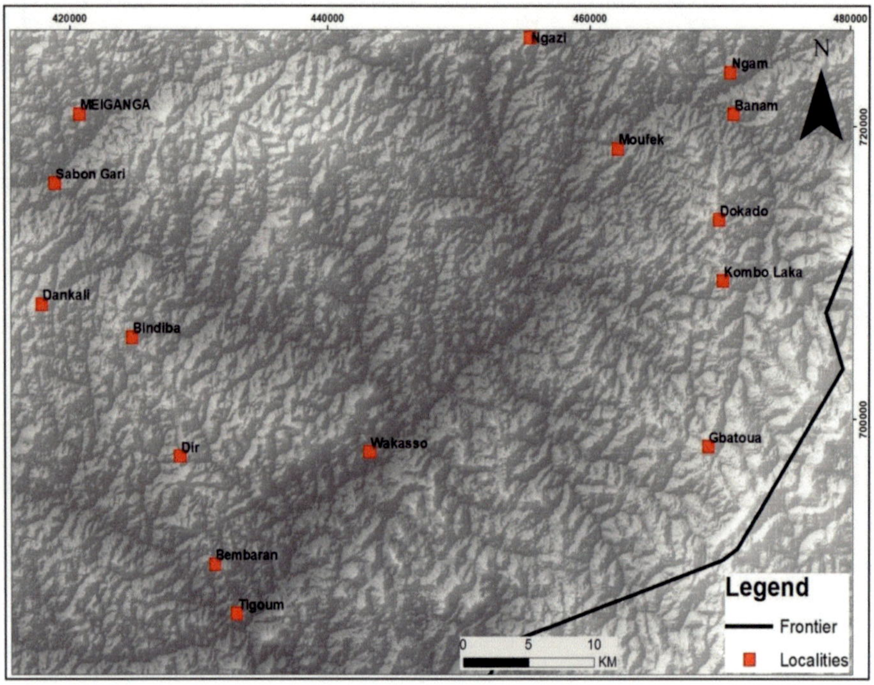

Figure 5.
Shaded relief of the zone derived from satellite image.

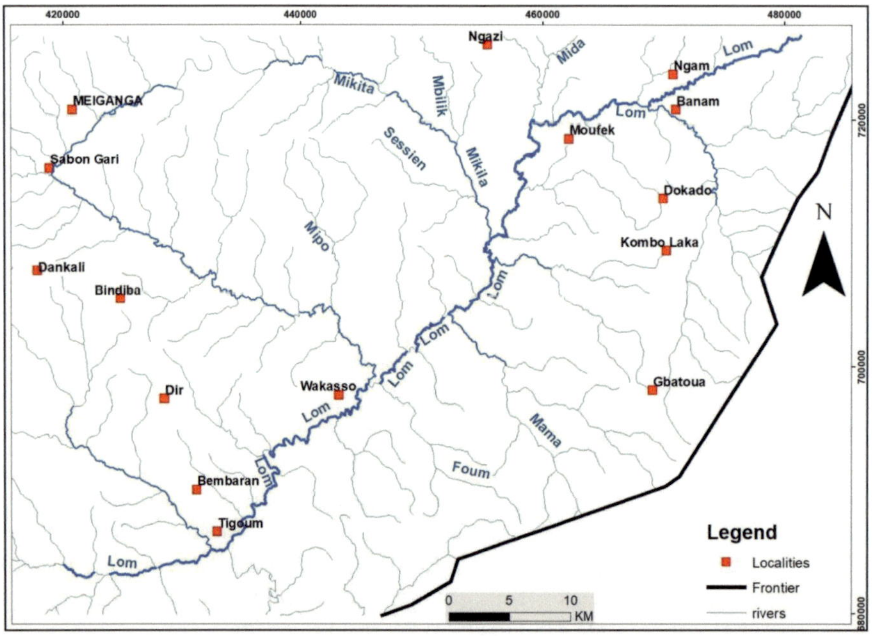

Figure 6.
Hydrography of the study area derived from the satellite map.

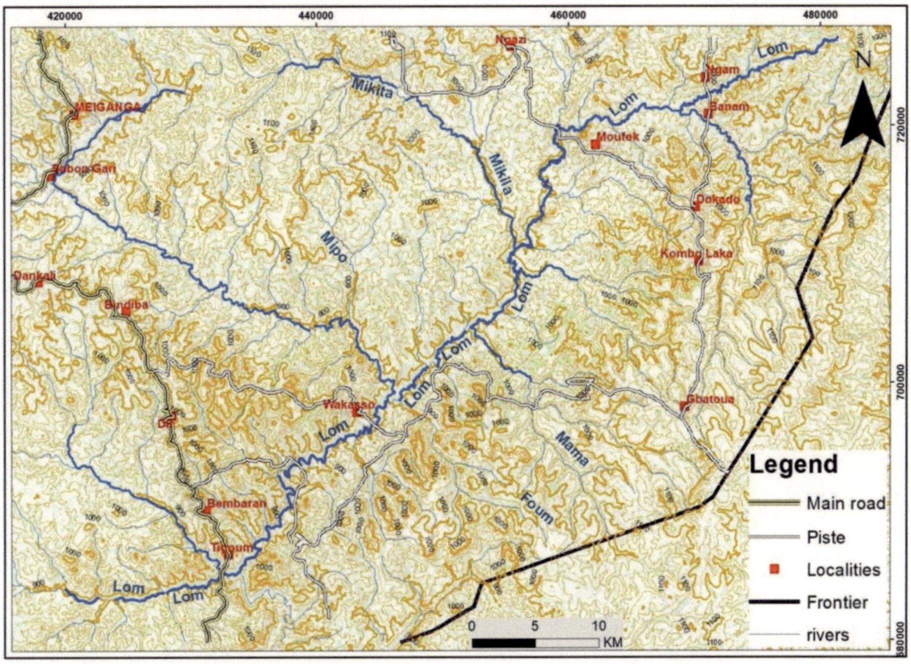

Figure 7.
Basemap of the Meiganga area showing contours, hydrography and localities constructed from the above maps.

Final Map: Planned points are points to be sampled on the field. The streams to be sampled are a part of the norms established by the project's needs (**Figure 8**) we see that the Lom river is flowing following the NW° SE direction.

3.2.3 Database and notebook

In Microsoft Access a database consists of one single file. The file contains all the tables of the database, the relationships, queries, forms (user windows), and many other things like:

- **The Standard operations procedures** which contains all the information related to how the work must be done on the field i.e. codes inherent to the project. A geologist named *Quinter Tanni* is attributed an initial like QT. Amphibolite and granite can be abbreviated like AMP and GRT respectively. Meiganga toposheet is named MEI (**Figure 9**). These codes help to secure information related to the project, for identification purposes, classification, retrieval and management of useful data.

- **The database** helps in the storage and management of all the information above including: geographical related data (maps and vectors) and non-related geographical data (codes, phones...).

At the end of the planning stage, a database is available inside the laptop of the geological team on the field to enter data collected. The database contains at this level only the planned points, the area where they are situated, and the name of the geologist authorized to enter the database (**Figure 9**). There are also codes for filling

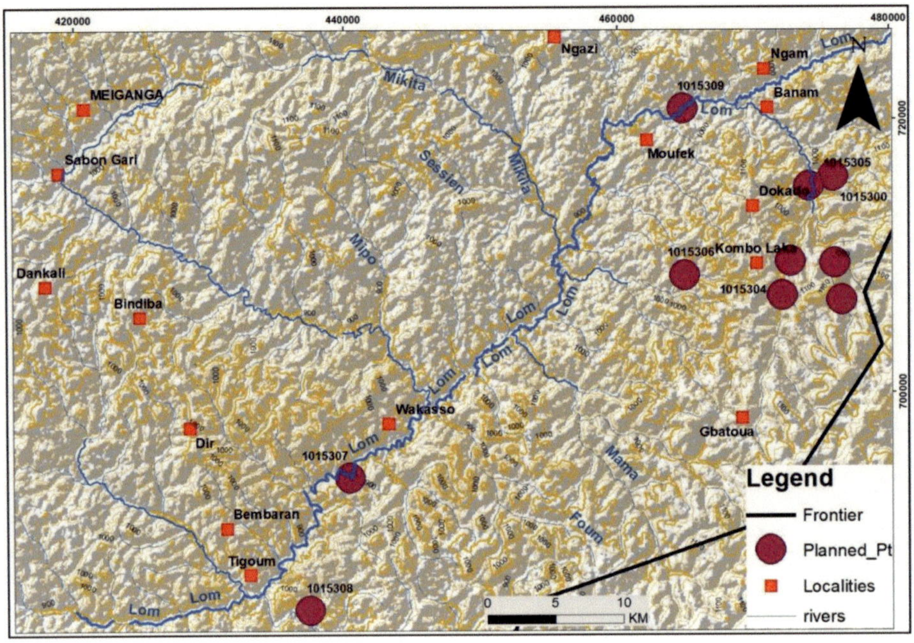

Figure 8.
Final map of the planning phase containing planned points, hydrography, contours, and forms of relief.

Employee_Initial	Employee_Name	Employee_Title
QT	TANNI Quinter	Geologist
JX	XXXXXX	Geologist
PY	XXXXXX	Geologist
JZ	XXXXXX	Supervisor

PROSPECT	
NAME	CODE
Meiganga	MEI
Poli	xxxxxxxx
Rey Bouba	xxxxxxxx
....	xxxxxxxx

ROCK TYPE		
Group	Description	Code
Sedimentary	Claystone	SCY
	....	xxxxxxxx
Metamorphic	Schist - Muscovite	SCHM
	Schist - Chlorite	SCCL
	Greenschist (undif	GST
	Quartzite	QZT
	Gneiss	GN
	Amphibolite	AMP
Igneous	Gabbro	MG
	....	xxxxxxxx
Others	....	xxxxxxxx

Figure 9.
Codes in use, built at the planning phase.

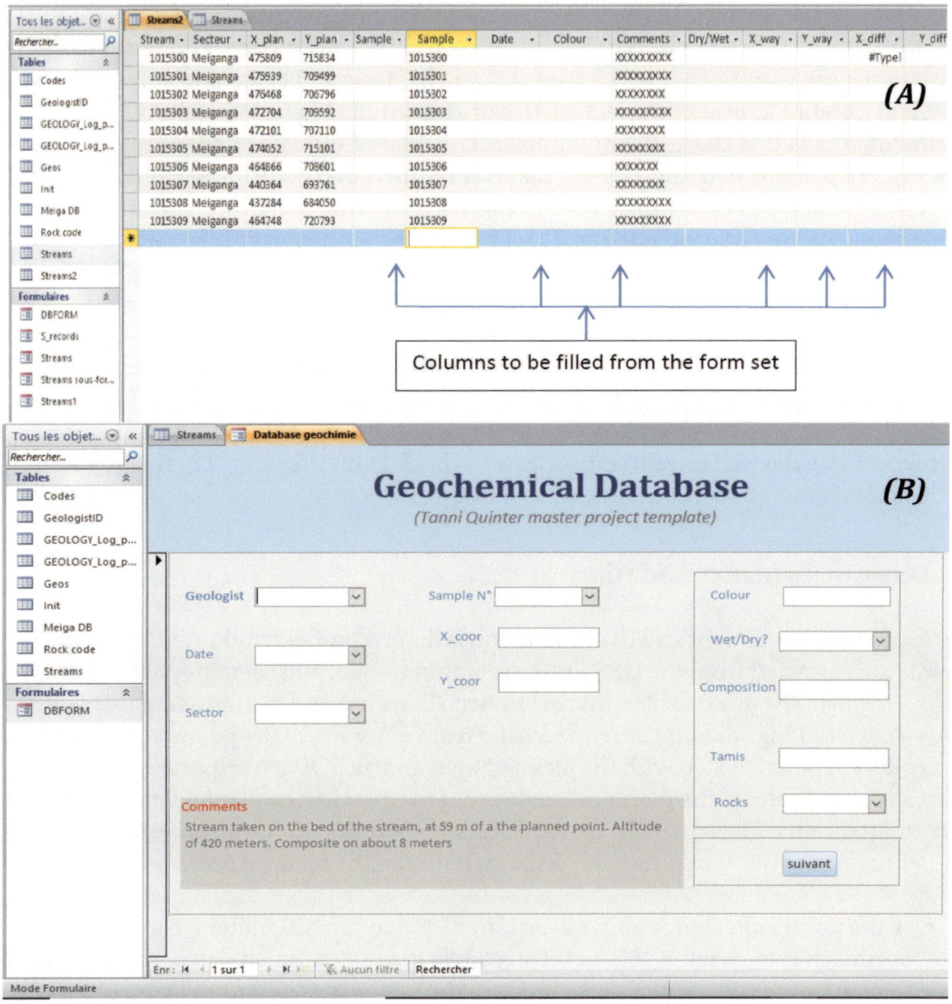

Figure 10.
(A) Database at the planning phase. (B) Empty form set at the planning phase.

the tables in the right way (**Figure 10A** and **B**). The user, a geologist who has to enter the data inside, access it through a form set. This used to be calling a soft database.

4. Field work/data storage

The field work consists of sampling the planned points and recording the associated data. The importance of GIS and database are with a great significance here. Effectively, the team on the field uses the planned data installed in the GPS and the database to collect the data and store them.

During the data sample collection phase, geologists use the GIS to plan their movements on the field, given that roads and topography are not uniform. With the GIS, a geologist can adopt the use of another road avoiding steep slopes. The GPS helps the geologist in navigation on the field. Inside the GPS, all the maps are inserted in order to help the geologists go through to their target. The GPS also contains pre-planned points done in the GIS. Also, the GPS ensures that the geologist is within a reasonable distance to sample the target, record the coordinates and mark the point with a specific name as defined in the SOP of the project. The geologist records inside his field notebook, others characteristics of the sample like color, texture, and environment etc. At this stage, the notebook is considered like the hard database. When the field trip is finished, all the data taken contained in the GPS and the notebook are recorded inside the database. Access can be used for such project. At this stage, it is consider being a soft database. Tracking and other data from the GPS are also extracted via Basecamp software capable to extract and send data from GPS. Those tracking and stored data will be called later in the GIS to assess the work achieved.

A map containing the area of interest is put into a GPS to ensure the safe navigation of a prospector on the field. In the case of the Meiganga topo sheet, pre-defined points are sampled following the GPS guiding and recorded (MARK) button to collect GPS coordinates on the sample points acting as a reference which will be further checked and validated once the prospector is back from the field. This exercise on the field is not complete without orderly labelling.

5. Assessment/data validation

Generally, assessment and data validation is done after fieldwork. All the sampled points are extracted from the GPS via Basecamp software, and then they are transformed in gpx format. The gpx can be imported in ArcGIS software. The first assessment done is the visual checking whereby the supervising team observes all the points and ensures that the collected points match with the planned ones and all SOP are respected before they are validated. If one of the points taken does not match with the planned one, then a particular attention is given to them for rectification. Two situations can arise here with the visual checking after measuring the distance between the planned and sampled points:

- If the point collected is above a certain distance say 100 meters, the best approach is to check if the target stream is the one sampled. Also, we can proceed to check other characteristics precisely the environment and to listen to the explanations of the prospector in order to validate the point.

- If the distance of 100 meters is not respected and the visual checking confirms that it's not the targeted stream that are been sampled, then the point is not valid.

After the visual assessment, the ACCESS database is also checked in order to verify that data entered here are in conformity to the information pertaining to the sampled point on ArcGIS. For example, no sample can be entered without their geologic characteristics.

6. Case study of the Meiganga area

6.1 Data and storage

6.1.1 Characteristics of stream sediments

Maps built from GIS help the geologist to navigate wisely towards the planned points and pick the sample. Those samples are packed inside plastic bags in order to be sent to the laboratory for analysis. Geological characteristics of the sample are then recorded inside the field note book and the database ready at the planning phase. Those characteristics are used further for interpretations. In the course of this work, two types of samples were taken i.e. stream sediment and rock sample:

- For stream sediments, five kilograms of the sample is required (**Figure 11A** and **B**). Although when the sample is wet, the weight can go up to seven. Apart of the coordinates, characteristics of the sample which have geological significance are recorded apart; those are color, composition, contamination (**Table 1**). Every sample receive a unique identity called ID like QTXXX (with QT = initial of Tanni Quinter; XXX = sample number planned).

Figure 11.
Field campaign: (A) picture showing sampling of stream sediments following preplanned sample points and (B) darkish brown color of stream sediment in a codified sample bag.

Sample	Characteristic	Color	Material	Contamination
Sample	Humid	Brown	Sandy	Organic
QT1015300	Humid	Dark gray	Intermediary	Village
QT1015301	Humid	Grayish	Clayey	Organic
QT1015302	Dry	Dark	Intermediary	Village
QT1015303	Wet	Grayish	Clayey	Organic
QT1015304	Dry	Brownish	Sandy	Organic
JX1015305	Humid	Brownish	Sandy	Road
PY1015306	Wet	Grayish	Clayey	Road
JX1015307	Wet	Dark gray	Intermediary	Organic
PY1015308	Dry	Brownish	Sandy	Gold panning

Table 1.
Characteristics of stream sediments.

Figure 12.
Field campaign investigations: (A) showing the geological description of a rock outcropping in flat and (B) darkish gray color of granite sample.

Sample	Grain size	Color	Wheateri	Name
RQT001	Medium	White gray	MW	Granite
RQT002	Coarsed	Gray pink	FR	Granite
RQT003	Coarsed	Gray pink	FR	Granite
RQT004	Coarsed	Gray pink	MW	Granite
RQT005	Medium	White gray	MW	Granite
RQT006	Medium	White gray	FR	Granite
RQT007	Medium	White gray	FR	Granite
RQT008	Coarsed	Gray pink	MW	Granite
RQT009	Medium	Gray pink	MW	Granite
RQT010			SW	Granite

Table 2.
Geological characteristics of rocks (MW = moderately weathered; SW = strongly weathered, FR = fresh rock).

- For rocks, a fresh part of the rock which can permit to make good geological observations like grain size matrix, color, alteration mineral is picked (**Figure 12A** and **B**). The sample here must also help to build tin section in case of interesting features inside like alterations minerals as sulfurs. Every sample receives a unique ID like RQTXXX (with R = initial for rock; QT = initial of Quinter Tanni; XXXX = sample successive rock number) **Table 2**.

The database is filled during this phase though a form set. **Figure 13** shows a form set we built as an example. A form set is a user interface of the database. Its importance is to avoid bad manipulation. The interface is built by the Database management team. The geologist is the user. The information primarily

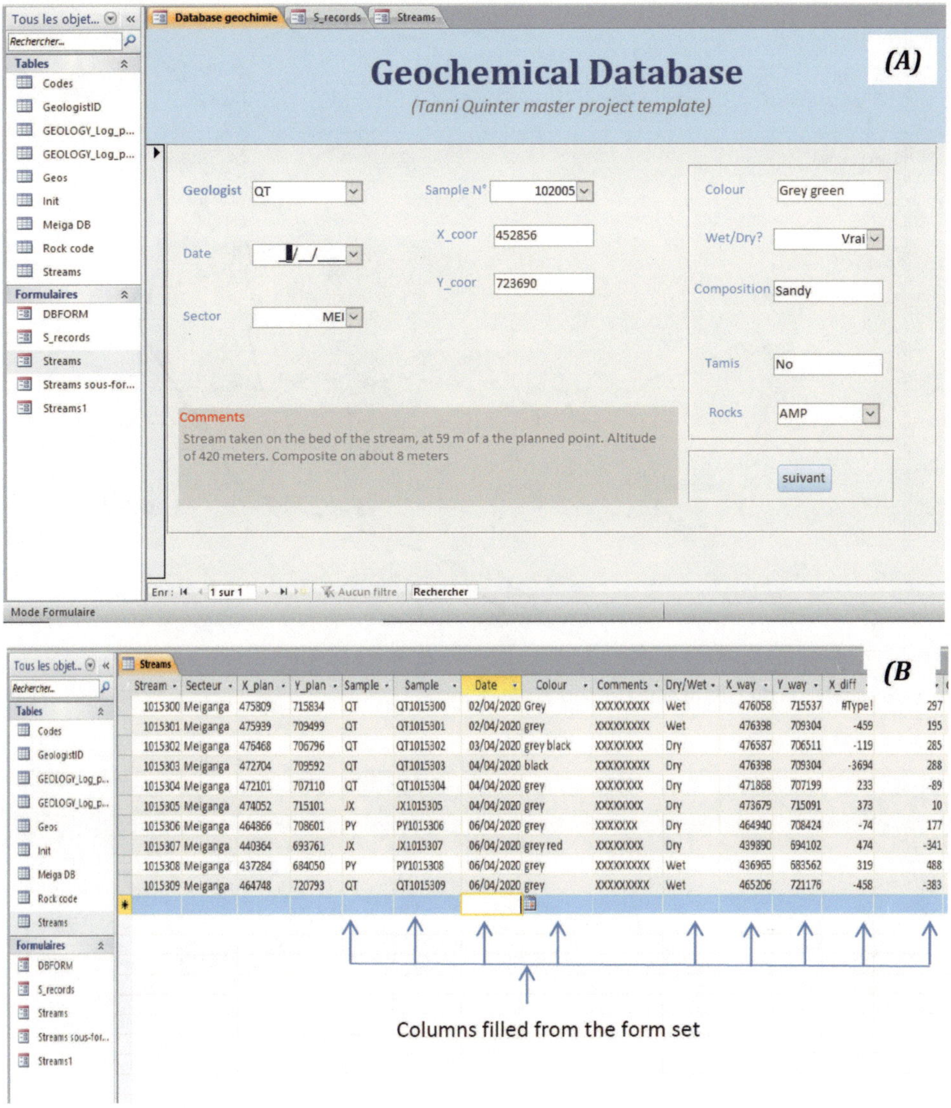

Figure 13.
(A, B) Filled form set built for the project. (B) Filled database.

recorded inside his notebook should be filled. The result is an integral database (**Figure 13**).

6.2 Tracking and waypoints

6.2.1 Tracking

During navigation, the geologist stores the road used inside his GPS through tracking (**Figure 14**). Tracking is recorded automatically. They show at which distance the

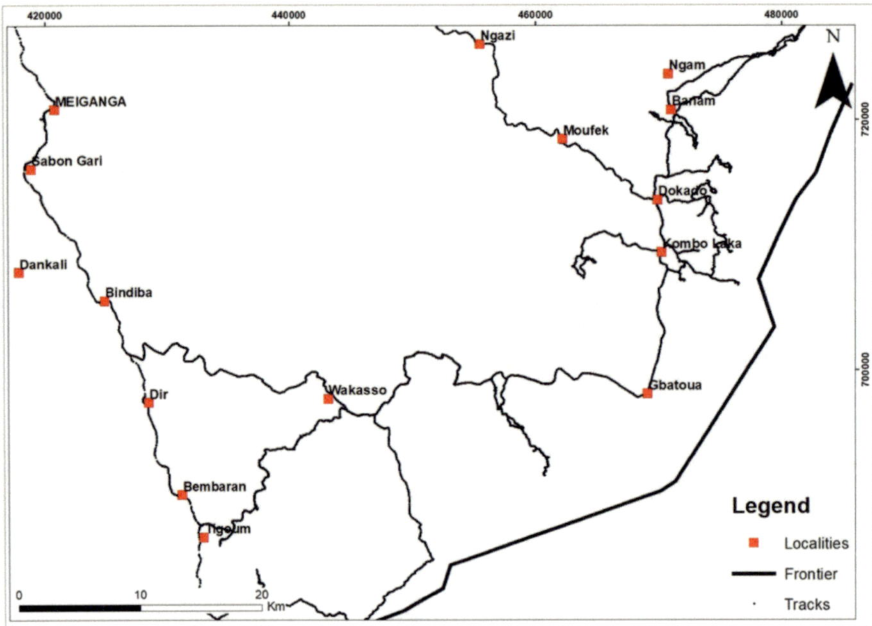

Figure 14.
Tracks extracted from the GPS of geologist on the field.

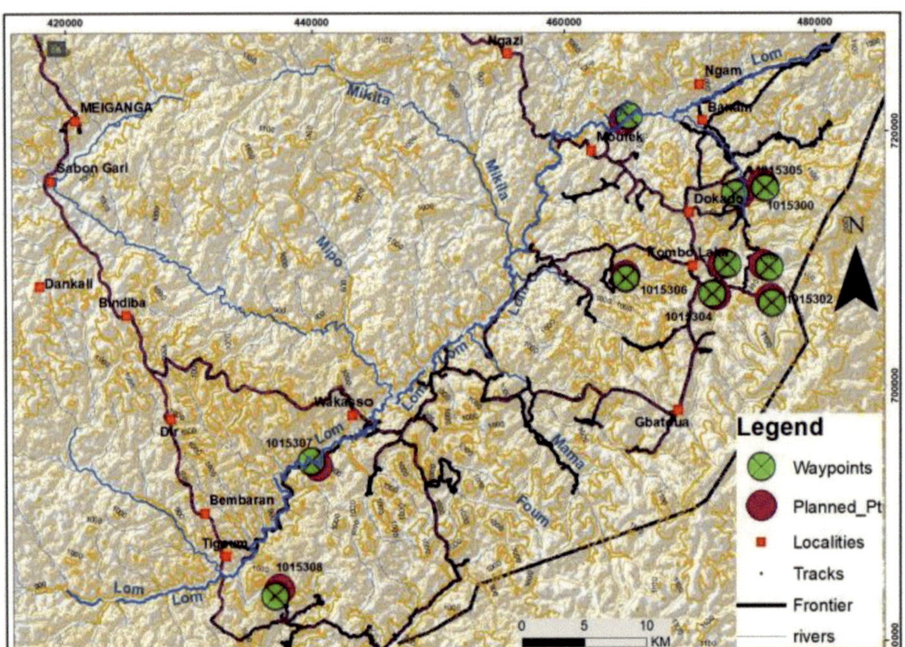

Figure 15.
Tracking and waypoints versus planned points (waypoints = stream sediments sampled).

geologist was from the point to sample. At the end of this stage, the tracking produced is extracted and stored in GPX format in Basecamp software. This will be used later for the assessment phase.

6.2.2 Stream sediment waypoints

Waypoints are points marked by the geologist while on the field (**Figure 15**). Every time a point is collected, a waypoint is marked inside the GPS, and coordinates of the points are recorded. Those waypoints are also extracted in GPX format and stored in Basecamp for assessment and validation.

6.2.3 Rock waypoints

Waypoints are also concerning rocks recorded on the field. Some are sampled and the others are just marked and observed. Those rocks information will help for the enrichment of the geological map (**Figure 16**). Although the information related to rocks are not assessed with the same importance as the stream sediments, they are also with great information and used later to link geochemical anomalies and the surrounding rocks formation. Since we know that the geochemical signature of the stream sampled bear the information of the rock where the sediments sampled are coming from. At this point, the tracking is used to assess if the geologist effectively reach the rock observed or sampled. Those waypoints are also extracted in gpx format and stored in Basecamp and inside the database.

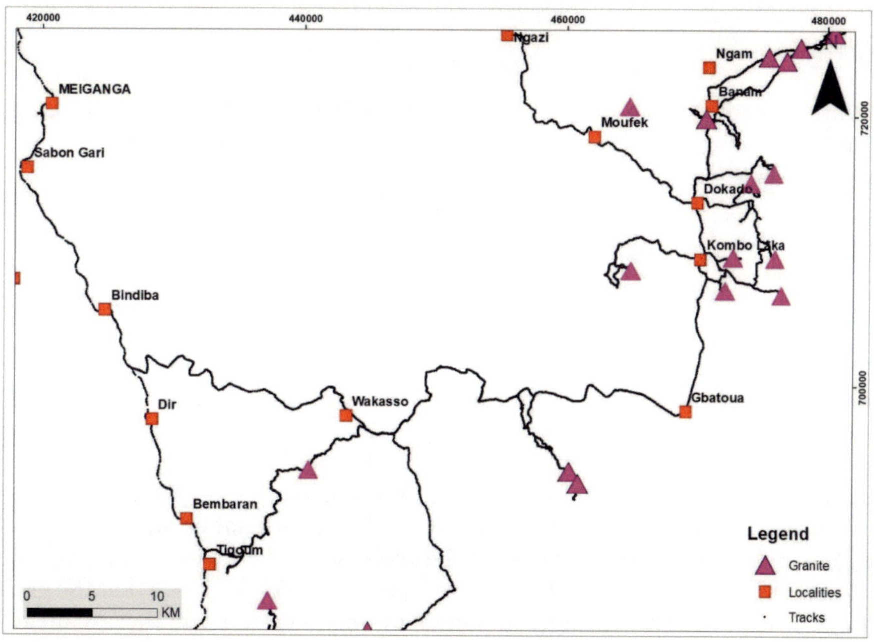

Figure 16.
Tracking and waypoints showing sampled and observed rocks (triangles = observed granites).

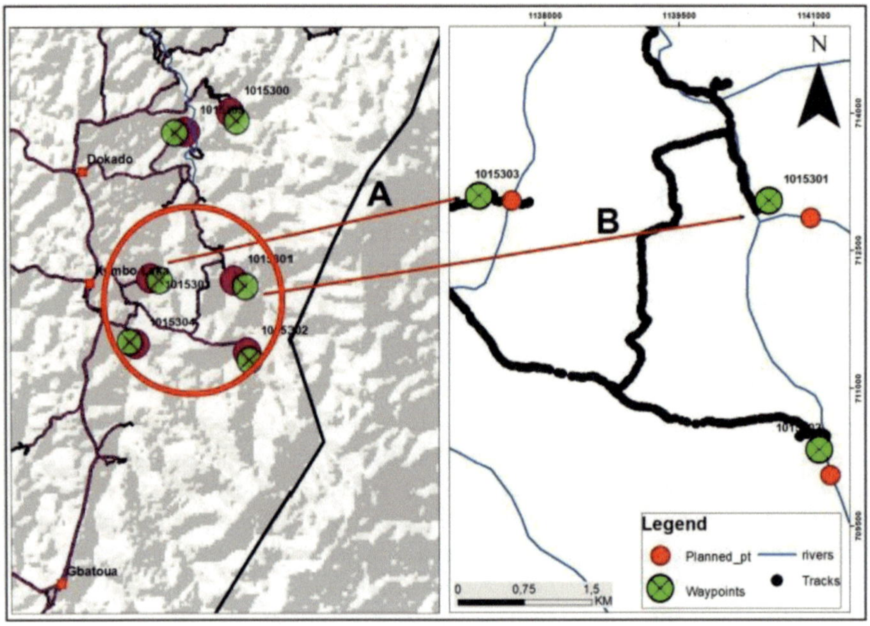

Figure 17.
Valid and invalid data after investigation.

7. Assessment and data validation

7.1 Valid data

The assessment process is made in two phases. The first is the visual checking and the second the parameters checking. According to the supervising team, at the end of this phase, data is valid when minimum criteria are met:

- Coordinates of sampled points correspond to those of planned points.
- Coordinates of the points sampled are within a distance of 100 m with the planned one, and situated on the same river.

The tracking of the geologist attests his movement towards the point planned to be sampled.

If a point does not respect the above criteria (**Figure 17**), an investigation is made in order to invalidate it effectively or not. Thus, the GIS is checked again to understand if the river has not shifted or if the targeted one does not exist. The database is assessed to understand the comments of the geologist.

Case A: Valid data after visual checking on the GIS. Tracks prove that the geologist effectively reach the targeted planned points. But he did not found the stream at that place. One of the reasons is that the river shifted at the left. So he came back where the stream crossed to sample. In that case, all the minimum criteria contained in the SOP (Standard operation procedures) are met.

Case B: Invalid data after visual checking on the GIS. Tracks prove that the geologist did not reach the targeted stream, or the targeted point. And the distance between

DOI: http://dx.doi.org/10.5772/intechopen.106632

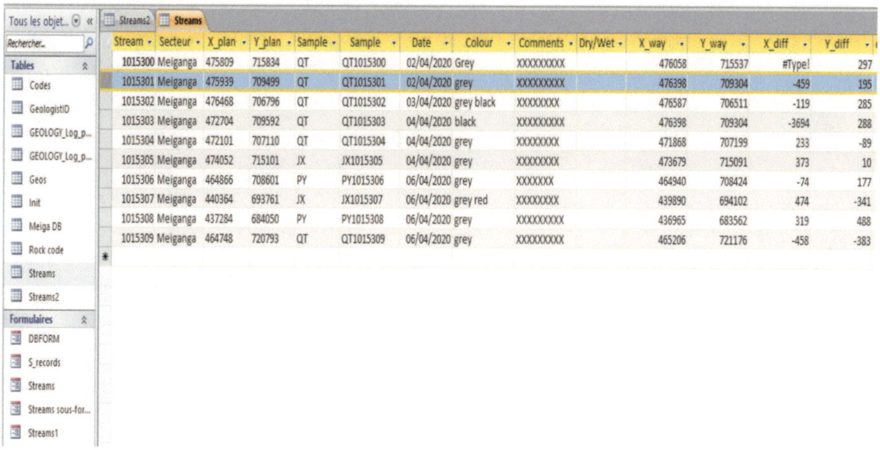

Stream	Secteur	X_plan	Y_plan	Sample	Sample	Date	Colour	Comments	Dry/Wet	X_way	Y_way	X_diff	Y_diff
1015300	Meiganga	475809	715834	QT	QT1015300	02/04/2020	Grey	XXXXXXXXX		476058	715537	#Type!	297
1015301	Meiganga	475939	709499	QT	QT1015301	02/04/2020	grey	XXXXXXXXX		476398	709304	-459	195
1015302	Meiganga	476468	706796	QT	QT1015302	03/04/2020	grey black	XXXXXXXX		476587	706511	-119	285
1015303	Meiganga	472704	709592	QT	QT1015303	04/04/2020	black	XXXXXXXXX		476398	709304	-3694	288
1015304	Meiganga	472101	707110	QT	QT1015304	04/04/2020	grey	XXXXXXXX		471868	707199	233	-89
1015305	Meiganga	474052	715101	JX	JX1015305	04/04/2020	grey	XXXXXXXX		473679	715091	373	10
1015306	Meiganga	464866	708601	PY	PY1015306	06/04/2020	grey	XXXXXXX		464940	708424	-74	177
1015307	Meiganga	440364	693761	JX	JX1015307	06/04/2020	grey red	XXXXXXXX		439890	694102	474	-341
1015308	Meiganga	437284	684050	PY	PY1015308	06/04/2020	grey	XXXXXXXXX		436965	683562	319	488
1015309	Meiganga	464748	720793	QT	QT1015309	06/04/2020	grey	XXXXXXXXX		465206	721176	-458	-383

Figure 18.
Checking of the database to see difference and comments. Note the X_diff case of −459 m between the planned and the waypoints recorded.

the waypoints and the planned one is >100 m. In that case, all the minimum criteria contained in the SOP (Standard operation procedures) are not met (**Figure 18**).

Data are declared invalid when they do not meet primary criteria of valid data, nor the additional ones. In that case, the point has to be collected again.

7.2 Validation of rock

Validation of rock sampled and observed is basically done by following the tracking of the geologists. If all the tracks cross the sampled and observed points, thus all the points are considered valid. In the following map, all the points are considered valid because the tracking confirms it to be so (**Figure 19**).

7.3 Contribution of our work

7.3.1 Data and their reliability

Following the work we achieved during the project, we learned that data taken on the field act as pointers used to determine whether or not the company will be a failure or a success. Attention needs to paid so as not to spend money to go back to get data which was not well collected. Thus, it is necessary to ensure as much as possible, that these data are representative and comprise significant geological information to understand them. Global positioning system coordinates of points are important. The conditions of samples and their characteristics are also important so that we can better comprehend for example whether a sample is contaminated.

Contamination for example significantly influences the obtainable quality of data. While an organic content is always present, gold can be present in a sample because of the gold panning site near like the one in Kombo Laka village at the east part of the map (**Figure 20**). During the assessment, it can be see that the village is up the sampled point and understand why an abnormal concentration of gold is present here and none in the samples near.

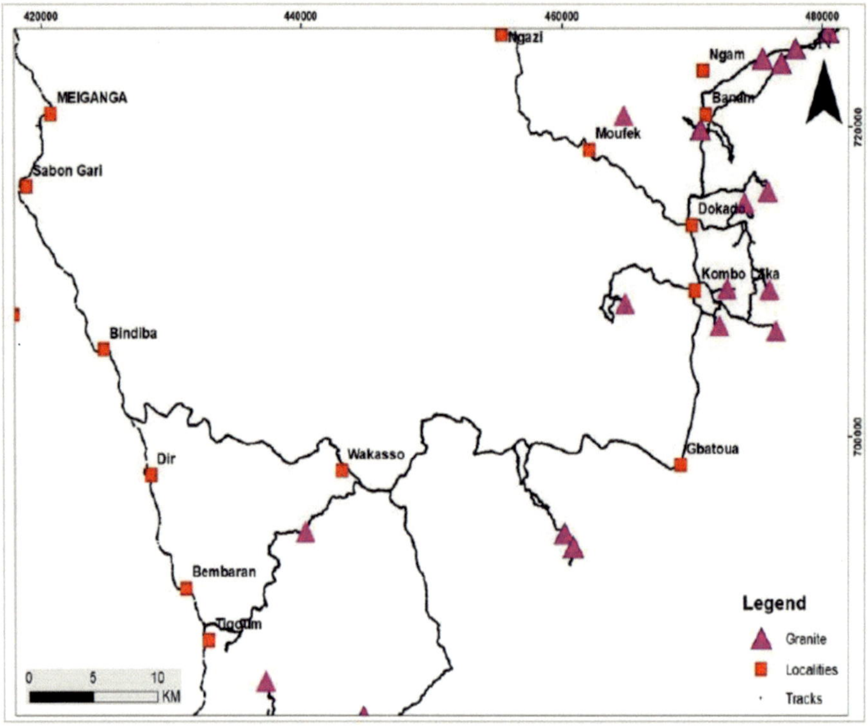

Figure 19.
Tracking versus sampled and observed rocks.

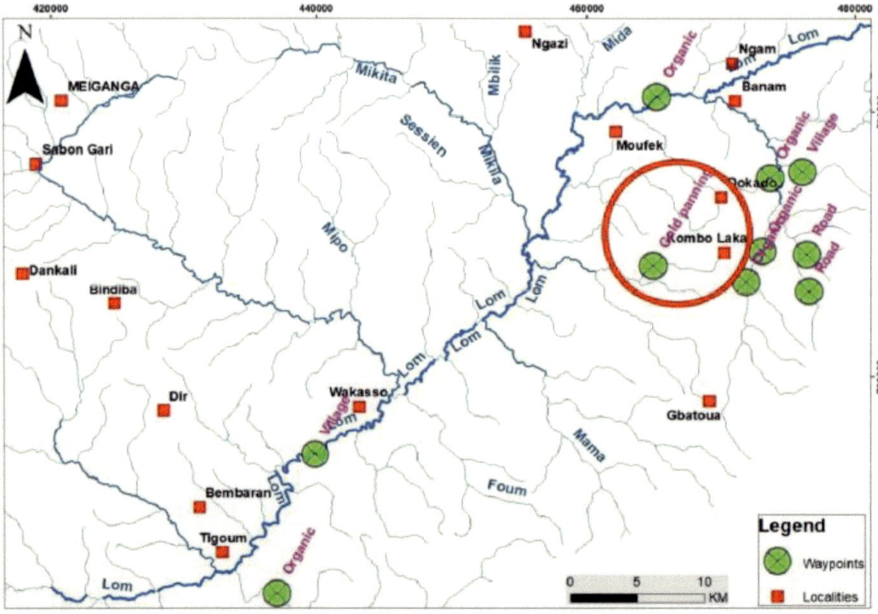

Figure 20.
Analysis of factors of contamination. The point is situated near the gold panning site.

Some tools that mining companies and mining projects use are Database and GIS. To use them effectively ask for skills and comprehension of the whole process to take reliable data than to discard some or retaken others.

7.3.2 Tools developed

The database has been effectively built in a software called Access and can allow the geologists of a mining company to record the data from the field without compromising the reliability of the database, this through the form sets. The same database permit to keep records that help for the assessment of data collected from the field. The database can be coded and adapted for any mining company and project. Inside every mining projects and company, a special team is dedicated to design, manage and secure the database.

Speaking about GIS, mining company can use the same hierarchy of our tools to take all the power of GIS and their importance at the very planning phase where all the data to be used on the field are collected and put together to allow a field team to deploy themselves safely. But also to assess the data they bring in order to validate them. Some data can be wrong. To avoid a mining company to rely on them and fall in failure, the data must be assessed and validated. Tracking and waypoints are used for that at this stage. But GIS and database also help further for interpretations of the data and their behavior.

7.3.3 Interpretation and sketch geological maps

When samples are taken, their geochemical anomaly helps to find ore bodies. Stream sediments is an exploration tool, which plays a significant role in geochemical exploration by identifying possible sources of anomalous element concentration. Sampling is done on rock debris, soil or other materials which has undergone erosion and transportation within a catchment basin upstream of a sampling site. It is thus representative of the geochemistry of materials from the upstream drainage basin. The collection and analysis of these stream sediments (silt, sand, mud, clay) in a stream or riverbed, either dry or wet are common methods for exploration. Stream sediments are used for geochemical analysis for enhancement of geological understanding.

Rock types can be differentiated by geochemical signatures in stream sediment data. But rocks can also help to draw sketch of geological maps. Those geological maps are then compared to the existing ones in other to understand how wrong were the map or not. The **Figure 21A** and **B** below shows a sketch of geological map built from the granites sampled in the Meiganga region.

7.4 Disadvantages of using GIS

Though GIS could benefit users from different disciplines it also has some disadvantages:

- *Expensive:* GIS setup is complex, in addition to the cost of the equipment and frequent updating of datasets or data models may lead to errors in results.

- *Real-time parameters*: The handling of growing datasets is an overall challenge to the GIS system.

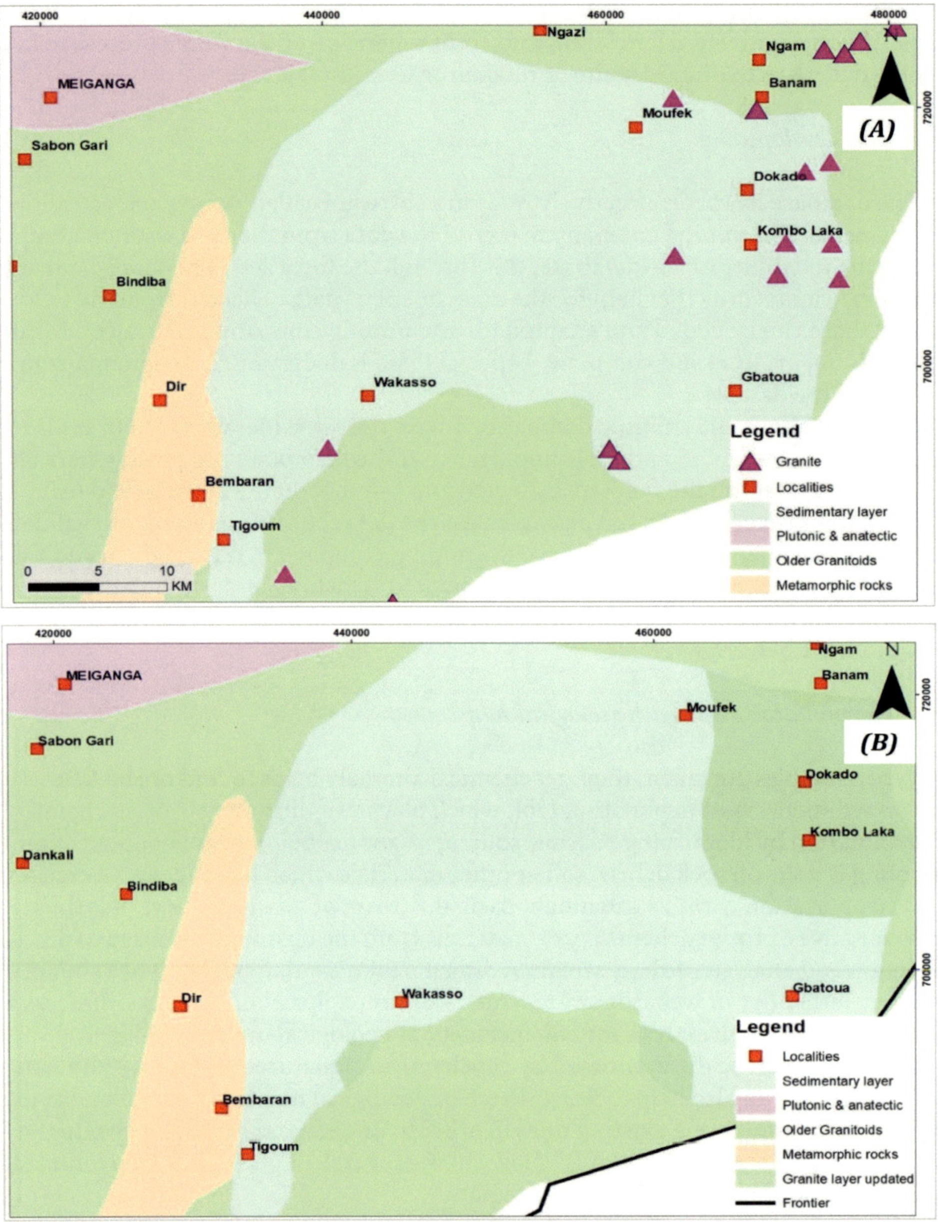

Figure 21.
(A, B) Sketch of geological map of the study area derived from sampled areas.

- *Geographical errors increase with larger-scale data*: The quality of the data collected directly affects the accuracy of the end system. Geographic errors will also affect net results since GIS handles large-scale data.

- *Relative loss of resolution*: Every technology has negotiable errors when deployed. In this Meiganga study area, some predefined points were displaced and did not fall on target streams (**Figure 22**).

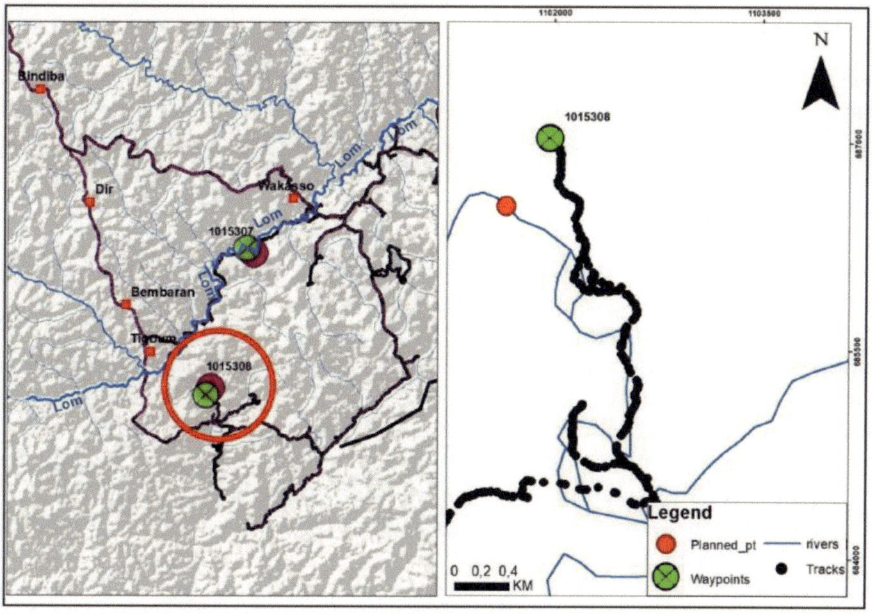

Figure 22.
Example of error. When zoomed, the point 1,015,308 which seems to be well placed on the stream, is shifted for more than 100 m.

- *Positional accuracy and precision*: Accuracy and precision are the functions of the scale at which a map (paper or digital) is created. The no spatial data linked to location may also be inaccurate or imprecise. Inaccuracies may result from mistakes of many sorts. No spatial data can also vary greatly in precision. Precise attribute information describes phenomena in great detail.

- *Violation of privacy*: The user community is not limited to authorized persons. So there is threat in the usage of data displayed from the GIS system.

- *Error-prone interpretation* could lead to failure of system implementation thus affecting the economic strategy of the implementer.

- *There might be failures in initiating or additional effort* required in order to fully implement the GIS but there might be large benefits to anticipate as well.

- *There is a lack of trained teachers in the domain*: Though GIS and remote sensing have been introduced in some universities, still the subjects have not yet been taught to the fullest extent. Moreover, a link between secondary education and higher education must be established for a wide spread and its continuity in the system. Prior knowledge of GIS is a prerequisite to train the trainers.

7.5 Improvements on GIS

GIS of the future may be very well done if already the following aspects are understood and objectives outlined following, other aspect such as: Planning, system design, implementation, maintenance of the software should be observed.

7.5.1 Planning

Your plan should answer the question: What information do I hope to gather? In planning its major contribution is to give us with an organized set of data which can help professionals to manage complex scenarios relating to the selection of site, environment impact, study of ecosystem, managing risk regarding the use of natural resources, sustainability issues, managing traffic congestion, routing of roads and pipelines etc.

Precise and accurate data is the core driving factor of any successful project. GIS is equipped with almost all those tools and functions that enables user to have access to the required data within a reasonable time.

Analysis is one of the major and most influential phases. Analysis guides us about the validity or correctness of design or we can say that analysis is a method which supports our design. Some of the analyses that can be performed by GIS are: water distribution analysis, traffic management analysis, soil analysis, site feasibility analysis, environment impact analysis, volume or area analysis of catchment, river or canals pattern analysis, temperature and humidity analysis, elevation etc. The construction phase involves the assemblage and putting into place the various tools to attain the objectives already established.

Lastly the operations phase, is the execution of a set of activities using all tools to acquire the necessary data. This last phase may, if not always necessitate a deployment of a qualified person to the field to obtain useful information.

8. Conclusion

At the end of this investigation which consisted of highlighting the importance of GIS and database in early stage of mining exploration, the objectives assigned, that were to build a reliable database and a GIS to ensure the planning phase is optimized and the results of this work meets standard operations procedures. The database built is capable to record and store geological and non-geographical information without being compromised by the mis-manipulation of geologists while at the same time it helps to ensure distances between sampled points and the planned ones do not exceed a certain distance.

The GIS is able to gather maps and geographical data for planning the field work and assessing results. Tools developed can then be used by any mining exploration project or company to work wisely and respect norms inside the mining sector. GIS and database are used in both Greenfield exploration projects to establish mining blocks based on statistics and in the brownfield exploration projects to do analysis using statistical methods like variograms to understand the behavior of our deposit. These analyses create grade control which guides in exploitation giving ideas of areas of high grade and low grade hence a suitable beneficiation method to be employed. The Meiganga area permitted us to effectuate the deployment and testing of our tools. However, some limitations have been highlighted concerning GIS. This shows that GIS and Database will always need to be in conformity with field work.

Acknowledgements

Part of the data presented here form a part of the second author's Master Thesis supervised by Dr. TAKODJOU WAMBO Jonas Didéro and M. NTOOGUE Paul Rene that was defended at the University of Yaoundé I, Cameroon. We gratefully acknowledge anonymous reviewers for their critical and constructive comments of the manuscript. We are thankful to Consulting and Training in Geosciences (CONTGEO) Research Centre for Mineral Exploration Targeting, for providing facilities during preparation of the earlier version of the manuscript.

Conflict of interest

The authors declare no conflict of interest.

Author details

Jonas Didero Takodjou Wambo[1,2]* and Wenzi Tanni Quinter[2]

1 California Institute of Technology, Pasadena, CA, USA

2 University of Yaounde I, Yaounde, Cameroon

*Address all correspondence to: jonasdidero@gmail.com

References

[1] Binam Mandeng EP, Bondjè Bidjeck LM, Takodjou Wambo JD, Taku A, Bineli Betsi T, Solange Ipan A, et al. Lithologic and structural mapping of the Abiete–Toko gold district in southern Cameroon, using Landsat 7 ETM+/SRTM. Comptes Rendus Geoscience. 2018;**350**(3):130-140

[2] Takodjou Wambo JD, Pour AB, Ganno S, Asimow PD, Zoheir B, dos Salles R, et al. Identifying high potential zones of gold mineralization in a sub-tropical region using Landsat-8 and ASTER remote sensing data: A case study of the Ngoura-Colomines goldfield, eastern Cameroon. Ore Geology Reviews. 2020;**122**:103530

[3] Takodjou Wambo JD, Ganno S, Djonthu Lahe YS, Kouankap Nono GD, Fossi DH, Tchouatcha MS, et al. Geostatistical and GIS analysis of the spatial variability of alluvial gold content in Ngoura-Colomines area, Eastern Cameroon: Implications for the exploration of primary gold deposit. Journal of African Earth Sciences. 2018;**142**:138-157

[4] Lisitsin V. Spatial data analysis of mineral deposit point patterns: Applications to exploration targeting. Ore Geology Reviews. 2015;**71**:861-881

[5] Gumiel P, Sanderson DJ, Arias M, Roberts S, Martín-Izard A. Analysis of the fractal clustering of ore deposits in the Spanish Iberian Pyrite Belt. Ore Geology Reviews. 2010;**38**(4):307-318

[6] Carlson CA. Spatial distribution of ore deposits. Geology. 1991;**19**(2):111-114

[7] Agterberg FP. Fractal modeling of mineral deposits. Proceedings 24th APCOM Symposium. 1993;**1**:43-53

[8] Porwal A, Carranza EJM. Introduction to the special issue: GIS-based mineral potential modelling and geological data analyses for mineral exploration. Ore Geology Reviews. 2015;**71**:477-483

[9] Carranza EJM, Hale M. Spatial association of mineral occurrences and curvilinear geological features. Mathematical Geology. 2002;**34**(2):203-221

[10] Carranza EJM, Mangaoang JC, Hale M. Application of mineral exploration models and GIS to generate mineral potential maps as input for optimum land-use planning in the Philippines. Natural Resources Research. 1999;**8**(2):165-173

[11] Asadi HH, Hale M. A predictive GIS model for mapping potential gold and base metal mineralization in Takab area, Iran. Computers & Geosciences. 2001;**27**(8):901-912

[12] Bonham-Carter GF, Agterberg FP. Arc-WofE: A GIS tool for statistical integration of mineral exploration datasets. Bulletin of the International Statistical Institute. 1999;**52**:497-500

[13] Setyadi H, Anggayana K. Database management and quality assurance is key of success in exploration. Procedia Earth and Planetary Science. 2013;**6**:42-49

[14] Setyadi H, Widodo LE, Notosiswoyo S, Saptawati P, Ismanto A, Hardjana I. GIS modeling using fuzzy logic approach in mineral prospecting based on geophysical data. AIP Conference Proceedings. 2016;**1711**(1):070002

[15] Fu C, Chen K, Yang Q, Chen J, Wang J, Liu J, et al. Mapping gold mineral prospectivity based on weights of evidence method in Southeast Asmara, Eritrea. Journal of African Earth Sciences. 2021;**176**:104143

Chapter 8

The Geomagnetic Field Transformants and Their Complexing with Data of Gravitational, Thermal and Radioactive Fields: During the Exploration of Hydrocarbon Fields at the Southern Part of the Ustyurt Region

Abetov Auez Egemberdievich,
Yessirkepova Sharbanu Bakhytovna and Julia Barbosa Curto Ma

Abstract

The chapter discusses the results of the interpretation of aeromagnetic survey data in the southern part of the Ustyurt region in order to identify zones that are heterogeneous in magnetic properties, with their subsequent tie-up with areas promising for the detection of hydrocarbon accumulations. Three tectonic elements are distinguished—the Central Ustyurt system of disturbances, the Shakhpakhty step and the Assakeaudan depression according to the sign, orientation and quantitative values of the transformants of the geomagnetic field, and, consequently, according to the depth and extent of distribution of magnetized rocks, based on the degree of deformation of lithological-stratigraphic complexes and levels by deep faults. Within these tectonic elements, the transformants of the geomagnetic field were comprehensively interpreted with the data of gravitational, thermal and radiochemical fields to increase the reliability of the results of geological interpretation. This made it possible to study the behavior and characteristics of faults, to draw conclusions about the depth of gravitational and magnetically disturbing masses, about the degree of geological heterogeneity of large geostructures, as well as to trace the nature of the manifestation of local structures in geophysical fields. The Central Ustyurt system of dislocation, the Shakhpakhty step and the Assakeaudan depression differ in varying degrees of active tectonics and differentiation by the degree of dislocation of oil and gas prospective objects, fragmentation by faults, by the extent of distribution of reservoir rocks and oil and gas source suite, conditions of conservation of hydrocarbon accumulations, directions and channels of probable hydrocarbon migration, and assumptions are made about the relationship of oil and gas prospective structures and explored hydrocarbon field with large deep faults. The author's studies of transformants of magnetic and gravitational fields, geothermal and aerogamma spectrometry data clearly indicate in favor of the prospects of oil and gas potential of local structures of the Shakhpakhty step: Utezhan, Kozhantai, Northern Kozhantai, Otynshy. In general, the results obtained will serve as a reliable basis for clarifying and detailing the geological and structural-formation models of the Southern Ustyurt, which can be used as the basis for the design of geological exploration for the exploration of new hydrocarbon field using expensive and time-consuming seismic exploration of CDP-3D and deep drilling.

Keywords: geomagnetic field transformants, airborne magnetic survey, magnetic anisotropy, high-frequency and low-frequency components, vertical and horizontal derivatives, Euler points, gravitational, thermal, and radiochemical fields

1. Introduction

1.1 Geological setting

The South Ustyurt region stands out on the southeastern margin of the South Mangyshlak sedimentary basin and is the eastern margin of the North Caucasian-Mangyshlak oil and gas province. This region is isolated as part of the South Mangyshlak oil and gas region, located within the Turan epihercynian platform.

In tectonic terms, the South Mangyshlak oil and gas region is located within a large area of Mesozoic intraplatform subsidence of the earth's crust, which is called the South Mangyshlak-Assakeaudan system of depression. This system is bounded in the north by the Mangyshlak-Central Ustyurt uplift zone, in the south by the northern periclines of the Karabogaz arch and the Tuarkyr uplift zone, and in the west, it opens toward the Caspian Sea.

This system of depression has the shape of a triangle in plan, with the apex to the west and the base to the east. The axial part of this zone is complicated by a linearly elongated chain of bath-like depressions separated by structural saddles.

These depressions are distinguished by different areas and irregular, most often isometric shape in plan.

Along the axis of the South Mangyshlak-Assakaudan system of depressions, the Zhazgurli, Uchkuduk, and Assakaudan troughs are differentiated from west to east. Within each of these depressions, smaller depressions are of particular importance, complicating their structure.

A characteristic feature of the tectonics of the South Mangyshlak-Assakaudan system depression is the stepwise subsidence of rocks from the boards of this system to its axial part. On the northern boards, the steps are located parallel or echelon to each other and generally subordinate the trend of the Mangyshlak-Central Ustyurt uplift zone.

Within the steps, chains of local uplifts of the anticlinal type are isolated, serving as traps for oil and gas. Local anticlinal folds are characterized by an asymmetric structure with a steep southern and more gently sloping northern wings and are slightly disturbed by faults.

The most ancient deposits, represented by thick strata of Upper Permian and Triassic rocks, are exposed in the areas of Mountain Mangyshlak. These rocks are drastically deformed and partially metamorphosed.

In the central and most submerged part of the South Mangyshlak oil and gas region, these occurrences are found at depths of more than 4 km. They are overlain by thick Jurassic sediments represented by all three sections, showing a drastic unconformity.

The stratigraphic completeness and thickness of these sediments within local tectonic elements are different. The Jurassic sediments in this area can reach a maximum thickness of 750 m. Traditionally, these sediments are associated with the main prospects for the exploration of hydrocarbon fields within the South Ustyurt. According to the conditions of sedimentation and the type of organic matter, the presence of main gas deposits is predicted here.

The rocks of the Jurassic complex with erosion and angular unconformity are overlain by deposits of the Cretaceous system, the eroded surface of which is overlain by Paleogene sediments.

Cretaceous deposits are widespread and are represented by the lower and upper sections.

Lower Cretaceous: The Neocomian complex includes sediments of the Barremian, Valanginian, and Hauterivian stages. At the bottom of the Neocomian stage, there is a layer of bluish-gray clay with a greenish tint from the Valanginian stage. This layer contains pyritized charred plant remains and thin interlayers of fine-grained sandstone. Above this layer, there is a sequence of interbedding sandstones, siltstones, and, greenish-gray clays with the inclusion of Hauterivian fauna.

The Barremian is the thickest in the Neocomian. The lower part of the Barremian deposits is composed of red-colored clays with subordinate interlayers of siltstones and sandstones; the upper one is predominantly sandstone.

Below the red-colored sediments of the Barremian stage lies a layer of gray-colored rocks, which in the upper part transition to variegated Aptian rocks. The complex is composed of dark gray, silty, finely elutriated clays with charred plant remains.

The Albian stage is represented by dark gray clays with a greenish tint, thinly bedded, silty, with charred plant remains and pyrite. Sandstones are gray and greenish-gray, mostly weakly cemented, often turning into loose sands, quartz-feldspar, glauconite.

The Upper Cretaceous is represented by the Cenomanian, Turonian, Senonian, and Danish stages. The Lower Cretaceous sediments are lithologically subdivided into two complexes: terrigenous rock assemblages (Cenomanian, Turonian) and carbonate rock assemblages (Senonian, Danish).

In general, the above sediments are represented by terrigenous varieties, with the exception of the Upper Turonian-Danish carbonate deposits and layers of limestone-shell rocks in the sandy-clay section of the Neogene.

Paleogene deposits with erosion lie on the Cretaceous deposits and are represented by three sections. The thickness of the Paleogene sediments ranges from 0 to 632 m.

Paleocene sediments are represented in the lower part by greenish-gray clayey limestones, and in the upper part by brownish marls.

Eocene sediments consist of greenish-gray marls, and in the upper part of the section, they transition to dense, strong brownish-gray marls. These sediments contain remains of fauna, plant detritus, and pyrite.

The section of the Oligocene age is composed of clays of marine origin, which conformably lie on Eocene deposits. The clays in this section are greenish-gray with a bluish tint, and are interbedded with siltstone and sands. They are calcareous, and contain faunal remains as well as pyrite.

Neogene sediments are widespread and commonly overlie Oligocene rocks, characterized by erosion and angular unconformity. The Neogene section is divided into two stages: the Tortonian and Sarmatian, which correspond in age to the Middle and Upper Miocene, respectively. Lithologically, the sediments are composed of marls and limestones, white, chalk-like, with inclusions of pyrite and gypsum. Sarmatian limestones make up the shell of the Ustyurt plateau. The thickness of the Neogene sediments ranges from 0 to 90 m.

The rocks of the Quaternary age lie on sediment deposits of different ages of the Cretaceous, Paleogene, and Neogene. The sediments are represented by brownish-yellow, quartz-feldspar sands and loose sandstones, siltstones, and gray loams. The thickness of the Quaternary deposits varies in the range of 0–15 m.

1.2 Data and materials

The relevance of the application of effective innovative methods in the exploration of hydrocarbon fields is currently growing due to the emerging trend of depletion of oil and gas fields in operation.

An urgent task is to develop algorithms for solving geological problems to replenish Kazakhstan's mineral resource base through the exploration and development of new hydrocarbon fields [1, 2].

In 2019, LLP "SPC GEOKEN" Company performed integrated geophysical study to identify oil and gas prospective areas at the Central Ustyurt system of dislocation and in the southeastern part of the South Mangyshlak-Ustyurt system of depressions to solve these problems.

An independent role in these studies was assigned to aeromagnetic exploration with the solution of specific geological problems.

An aeromagnetic survey was conducted using the GT-MAG-2 airborne geophysical system, with flight lines spaced 1000 m apart and flown at an altitude of 100 m above the ground surface. The daily variations of the magnetic field were taken into account by the PBM Pico complex based on a CS-3 cesium magnetometer.

The TrimbleR7 GNSS L1/L2 dual-frequency GPS receiver was used as a ground reference station for geodetic referencing. The alignment accuracy of the survey route in plan was ±1 m.

The field processing of aeromagnetic survey data was performed in two stages.

At the first stage, field processing was carried out, during which correction procedure was made for the deviation of instruments, daily variations in the magnetic field, and tie-up of survey lines using the statistical equalization method.

At the second, cameral stage, the following procedures were performed:

a. calculation and input of corrections for radio-altimeter's readings and hypsometry of the day surface topography;

b. filtering and correction of aeromagnetometry data;

c. compiling of a base of magnetic variation data;

d. calculation of normal and anomalous geomagnetic field and its most informative transformants;

e. geomagnetic field transformants mapping and profiles plotting.

In order to extract the maximum geological information, the following transformants of the observed geomagnetic field were calculated [3]: analytical signal, vertical and horizontal gradients of the anomalous magnetic field, the angle of the magnetic field gradient vector, high-frequency and low-frequency components of the magnetic field, autotracing of the anomaly axes ΔTa, position of singular points Euler (**Table 1**).

Potential fields data	**Transformants**	**Tectonic elements**		
		Assakeaudan depression	**Shakhpakhty fault terrace**	**Central Ustyurt uplift**
Geomagnetic anomalies field	Anomalous magnetic field modified to the pole, nT.	From 100 to 300	From −50 to 100	From 100 to 400
	Analytic signal, nT/m.	From 0 to 0.007	From 0.005 to 0.01	From 0.003 to 0.03
	Tilt angle of TDR transformant, radian.	From 0 to 1.575	From −1.570 to 0	From −1.250 to 1.600
	Vertical derivative of geomagnetic field anomalies, nT/m	From −0.003 to 0.007	From 0 to −0.007	From −0.02 to 0.02
	Horizontal derivative of geomagnetic field anomalies, nT/m.	From 0 to 0.007	From 0 to 0.003	From 0.002 to 0.015
	Anisotropic transformant of the anomalous geomagnetic field, cu.	From 0 to 0.0025	From 0 to 0.0007	From 0 to 0.003
	Autotracing of axes of the anomalous geomagnetic field, cu.	From −0.7 to 1.5	From −3 to 0.7	From −3 to 6
Gravity anomalies field	Module of horizontal derivate of Bouguer anomaly, mGl/m.	From 0 to 0.0015	From 0 to 0.0015	From 0 to 0.0033
	Module of the vertical derivate of Bouguer anomaly, mGl/m.	From −0.002 to 0.001	From −0.001 to 0.001	From −0.003 to 0.004
	Local gravity anomaly (recalculated in the upper half-space at a height of 2.5 km), mGl.	From −3 to 1	From −1.5 to 3	From −2 to 8
	Anisotropy transform, the accent of maximum gravity anomaly. Sliding window size 5 km × 2 km, cu.	From 0.1 to 0.46	From 0.1 to 0.48	From 0.1 to 0.50
Thermal anomalies field	Thermal field according to satellite imagery.	From 6162.9 to 6713.3	From 6106.4 to 6608.5	From 6237.7 to 6462.4
	Map of the regional component of the thermal field.	From 6297.8 to 6601.5	From 6283.1 to 6543.4	From 6319.8 to 6395
	Map of the local component of the thermal field.	From −182.2 to 152.8	From −211.2 to 64.8	From −65 to 69
	Map of the thermal field recalculated in the lower half-space. A cut at 1500 m.	From −17.2 to 16.6	From −17.9 to 12.6	From −10.5 to 12.6
	Map of the thermal field recalculated in the lower half-space. A cut at 2500 m.	From −20.8 to 20.9	From −22.2 to 12.1	From −8.9 to 9.8

Potential fields data	Transformants	Tectonic elements		
		Assakeaudan depression	Shakhpakhty fault terrace	Central Ustyurt uplift
Radiogeochemical anomalies field	Potassium value (K), %	From 0.5 to 1.3	From 0.4 to 1.3	From 0.6 to 1.3
	Thorium value (Th), ·10–4%	From 1.4 to 4.0	From 1.4 to 4.0	From 1.5 to 4.0
	Uranium value (U), ·10–4%	From 1.5 to 3.5	From 1.0 to 3.0	From 1.1 to 3.1
	Dose rate gamma radioactivity (DR)	From 16.6 to 38.2	From 13.9 to 36.5	From 16.7 to 37.2
	U/K, cu	From 1.7 to 4.1	From 1.6 to 3.7	From 1.7 to 3.5
	K/Th, cu	From 2.1 to 3.3	From 2.4 to 3.3	From 2.5 to 3.3
	U/Th, cu	From 0.6 to 2.0	From 0.7 to 2.3	From 0.7 to 2.0

Table 1.
Statistics on the values of transforms of the anomalous magnetic and gravitational, thermal, and radiogeochemical fields.

In the Geosoft Oasis Montaj software, matrices were built (with a cell size of 250 × 250 m) using the "Bidirectional Line Gridding" algorithm.

The international analytical model IGRF was adopted as the normal geomagnetic field of the Earth.

2. Results of the geological interpretation of the geomagnetic field transformants

2.1 Structure of the observed geomagnetic field

The poleward geomagnetic field in the study region is represented by linearly elongated ΔTa anomalies of both signs and is characterized by a northwest trend (**Figure 1A**).

The Karabaur swell is characterized by high values of the geomagnetic field intensity (+240 + 360 nT, in places up to +520 + 800 nT) with maximum values on the northern periphery of this swell, which indirectly indicates its dipping in the northern direction under the North Ustyurt massif.

The southern periphery of the Central Ustyurt system of dislocations is marked by a drastic decrease in the magnetization of rocks, which is demonstrated by the minimum values of the geomagnetic field intensity (up to ± 50 nT).

The boundary separating the Shakhpakhty step and the Central Ustyurt system of dislocations (Kolsai trough and Koskudyk swell) is distinguished by the minimum values of ΔTa anomalies (from +46 nT to +14 nT).

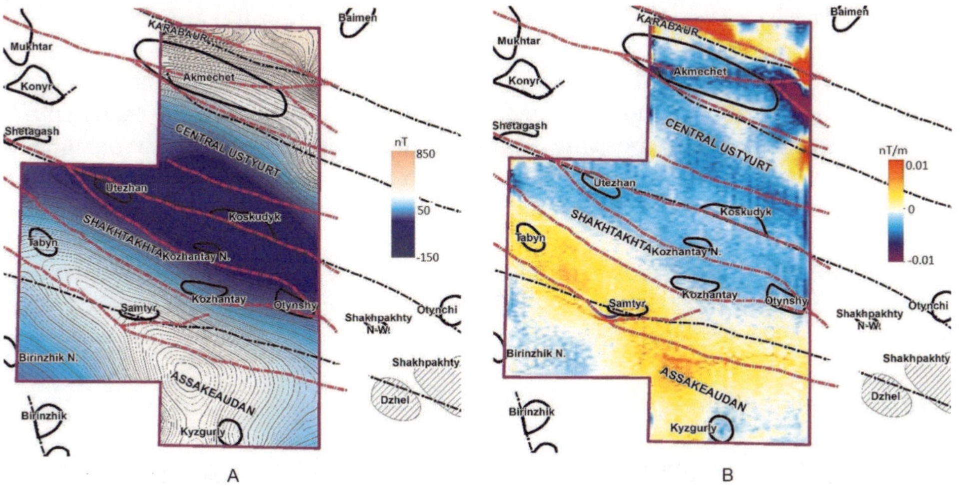

Figure 1.
(A) Anomalous magnetic field reduced-to-the-pole; (B): Vertical derivative of the anomalous magnetic field reduced-to-the-pole. The dashed black lines indicate major faults along V reflecting horizon. Dashed pink lines indicate major tectonic break highlighted on a set of completed research (gravity exploration, magnetic prospecting, thermal fields, and ground surface).

The northern part of the Shakhpakhty tectonic step is characterized by a negative anomalous geomagnetic field (up to −50-100 nT) on the local structures Utezhan, Koskudyk, Kozhantai, Northern Kozhantai, and Otynshy (**Figure 1A**).

This circumstance indicates the formation of an anomalous geomagnetic field at the northern part of the Shakhpakhty step at a different time than at the Central Ustyurt fault system of dislocation and the Assakeaudan depression.

At the southern part of the Shakhpakhty step, another reversal of the geomagnetic field was revealed. Here, a high-gradient zone of positive anomalies ΔTa manifests itself with quantitative values up to +100 + 120 nT (**Figure 1A**).

At the Assakeaudan depression, large anomalies of the geomagnetic field of northwestern orientation with an intensity of up to +300 nT have been established (**Figure 1**).

2.2 Vertical gradient of the geomagnetic field

The tectonic boundaries between the Central Ustyurt system of dislocation, the northern and southern parts of the Shakhpakhty tectonic step, and the Assakeaudan trough show themselves sharply and rather contrastingly in the transformant (dZ) of the geomagnetic field (**Figure 1B**).

The Assakeaudan depression is characterized by minimal variations in this transformant (−0.002 + 0.006 nT/m), while the Shakhpakhty step exhibits intermediate values (0 to −0.006 nT/m). The maximum values are observed at the Central Ustyurt system of dislocations (−0.01 + 0.01 nT/m) (**Table 1**).

The vertical gradient of the geomagnetic field localizes anomalies, reveals the block structure of magnetic and non-magnetic rocks, emphasizes the high-frequency component, and makes it possible to see more clearly the structural and tectonic features along the chains of anomalies when mapping faults. Presumably, some of them were migration of hydrocarbons in the presence of favorable structural and tectonic conditions.

If traps were encountered along the paths of migrating hydrocarbons, then the possibility of their filling is not excluded. In these cases, changes in the redox environment above the field are possible, leading to the formation of such ferromagnets as magnetite, hematite, pyrrhotite, which could be reflected in the structure of the high-frequency component of the geomagnetic field in the form of dissected and broken isodynamic lines.

Presumably, such processes occurred on the Utezhan, Kozhantai, Northern Kozhantai local structures, which is confirmed in the field of negative values of this transformant. However, this geological phenomenon needs further study.

2.3 Horizontal gradient of the geomagnetic field

The total horizontal poleward gradient was calculated by analyzing the lateral variability of the geomagnetic field (**Figure 2A**). The Shakhpakhty tectonic step exhibits intermediate values of the gradient of this transformant (from 0 to 0.002 nT/m), which increase to the north up to 0.010 nT/m, in the direction of the Central Ustyurt system of dislocations. Conversely, in the south direction, toward the Assakeaudan depression, the gradient decreases to 0 to 0.006 nT/m (**Table 1**). These findings directly confirm the presence of lateral inhomogeneities of magnetically disturbing masses in the study region.

2.4 Magnetic anisotropy

The magnetic anisotropy transformant, reduced to the pole (**Figure 2B**), shows the boundaries of anomaly-forming objects in contrast. To calculate this transformant, a sliding window size of 5 x 2 km was used.

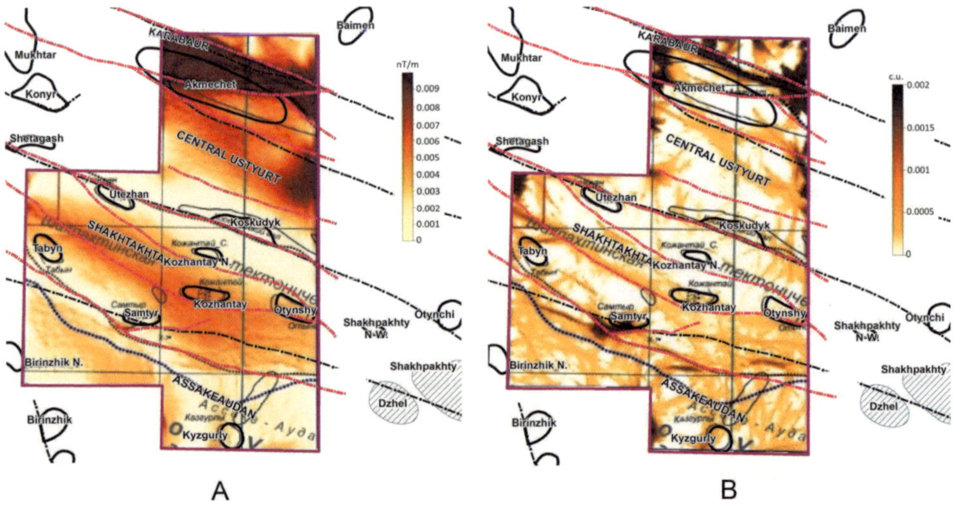

Figure 2.
(A) Module of the horizontal gradient of the geomagnetic field normalized to the pole. (B): Anisotropic transformant of the geomagnetic field normalized to the pole. Black dotted lines indicate major faults. Dashed pink lines indicate tectonic faults identified by the complex of studies performed (gravity exploration, magnetic prospecting, thermal fields, and day surface topography).

The southern and northern boundaries of the Shakhpakhty tectonic step, in the zone of its junction with the Assakeaudan depression and the Central Ustyurt system of dislocations (up to 0.002 standard units), are characterized by higher-than-usual values of the magnetic anisotropy transformant, which indicates an increase in the degree of heterogeneity of the basement rocks (**Table 1**).

Actually, the Shakhpakhty step is distinguished by reduced values of magnetic anisotropy (0–0.0005 standard units), including the Utezhan, Kozhantai, Northen Kozhantai, Otynshy local structures (**Figure 2B**), which is considered as a favorable historical and geological factor that had an indirect sedimentary section.

2.5 Analytic signal

Several areas are distinguished according to the nature of the distribution of this transformant.

The areas of local structures Utezhan, Koskudyk, Kozhantai, Northen Kozhantai, and Otynshy are characterized by lower values of the analytical signal, which are isolated in the northern part of the Shakhpakhty tectonic step and in the zone of its junction with the Central Ustyurt dislocation system.

To the southeast of these local structures within the Shakhpakhty tectonic step, but situated in the territory of the Republic of Uzbekistan, Shakhpakhty and Dzhel large gas fields were explored in the Upper Jurassic sediments.

To the south of the Shakhpakhty step, within the Assakeaudan trough, the transformant of the analytical signal exhibits the minimum values at the Kyzgurly, Birinzhik, and Northern Birinzhik local structures.

At the Central Ustyurt system of dislocations (Akmechet structure), increased values of this transformant are observed.

The revealed zonation is quantitatively confirmed by the values of the transformant of the analytical signal. At the Shakhpakhty step, they vary in the range of 0.005–0.008 nT/m, at the Assakeaudan depression they are 0–0.005 nT/m and at the Central Ustyurt system of dislocations they are 0.002–0.02 nT/m (**Table 1**).

A similar nature of the distribution of this transformant, apparently, is associated with: a) the general deepening of the upper edges of the magnetically disturbing masses in a southerly direction; and b) the block structure of the basement, in which magnetically active rocks of various scales are distributed.

2.6 The angle of the geomagnetic field gradient vector—TDR (tilt derivative)

This transformant (measured in radians) shows the maxima of the initial geomagnetic field regardless of their intensity and makes it possible to trace structural elements and map the position of the studied objects and their contours [4, 5].

The assessment of the depth of the studied objects is determined by the intensity of the TDR maxima (of both signs), which correspond to the axial lines of magnetically active objects. Zero values are their outer boundaries.

The northern part of the Shakhpakhty tectonic step is distinguished by the minimum values of TDR transformant (from −1.567 to 0 radians) normalized to the pole (Utezhan, Koskudyk, Kozhantai, Northen Kozhantai, and Otynshy local structures) (**Table 1**).

Near values of this transformant (0–1.560 rad.) were recorded in the Assakeaudan depression and at the southern part of the Shakhpakhty step (Samtyr, Tabyn, Kyzgurly, Birinzhik, and Birinzhik northern structures).

The maximum negative values (−1.200–1.560 rad.) of the dip angles of the TDR transform (Akmechet structure) were found in the Central Ustyurt dislocation system.

2.7 Autotracing of axes of magnetic field anomalies normalized to the pole

This transformant is used to localize the axes of anomalies ΔTa (**Figure 3**) in order to identify faults, the presence of which is determined: a) by a sharp change in the intensity of the magnetic field along certain lines and b) by characteristic shifts of anomalous values relative to this line [6].

The axes of key anomalies help to distinguish the margin zone of Assakeaudan Depression, where the contact of which with Shakhpakhty Tectonic Step is revealed by a narrow boundary.

Minimal variations were observed in the auto-tracing of the axes of the geomagnetic field anomalies, which were normalized to the pole, in the Assakeaudan depression (−0.5 + 1 standard units). In contrast, intermediate values were found at the Shakhpakhty step (+0.5 to −3 standard units). The maximum values were established within the Central Ustyurt system of dislocations (±5 standard units), which is probably due to the involvement of this regional structure in the Late Paleozoic tectogenesis (**Figure 3**).

2.8 Depth distribution of special Euler's points

The depth distribution of magnetically disturbing masses is well illustrated by the Euler deconvolution solutions calculated in the Oasis Montaj software. We used a structural index of 1 with a window size of 20 km^2.

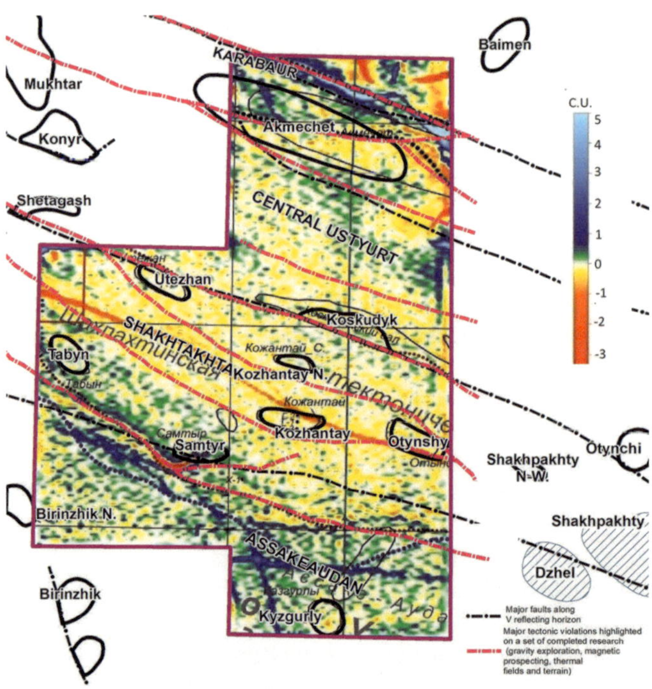

Figure 3.
Tracing of axes of magnetic field anomaly, normalized to the pole.

Clusters of Euler points indicate an increased density of magnetic anomalies and their grouping into ordered lines can be used to trace the depths and contours of anomaly-forming objects or magnetically disturbing masses [7].

The schemes of the distribution of Euler points on depth and density (**Figures 4–6**) and the structure diagram of these points in geomagnetic fields (**Figure 7**) convincingly testify in favor of the inhomogeneous magnetization of rocks occurring at different depths, which is also indicated by the previously considered transformants of the geomagnetic fields.

The distribution of magnetically disturbing masses in the study area is sharply differentiated on area and depth. For example, in the depth interval of 6–11 km in the northern part of the Shakhpakhta tectonic step, the Euler points are grouped into NE-trending bands (**Figure 5**).

Their position is not controlled by the fault. From which we can conclude that the faults identified here do not have "deep roots."

At the southern part of the Shakhpakhty tectonic step and at the Assakeaudan depression, Euler points form clouds of variable orientations from sublatitudinal to submeridional (in the depth interval 6–11 km), which call into doubt the tying between the genesis of magnetically disturbing masses and fault tectonics.

An indirectly revealed fact indicates a large total thickness of the rocks of the sedimentary cover and the quasi-platform cover on the Shakhpakhty tectonic step, which in turn indicates this tectonic element as favorable from the point of view of their oil and gas potential.

And finally, at the Central Ustyurt system of dislocations, the Euler points are grouped in the form of a zone of northwest orientation in the depth interval of

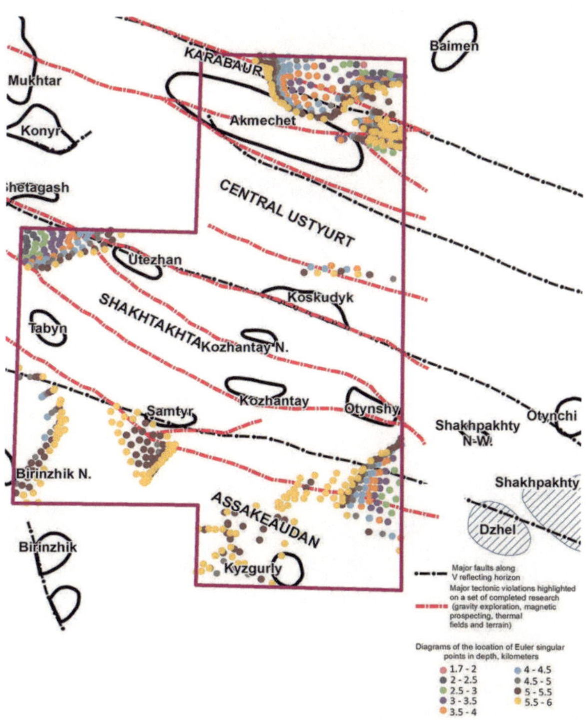

Figure 4.
Scheme of distribution by depth and density of Euler points (up to 6 km) of the geomagnetic field.

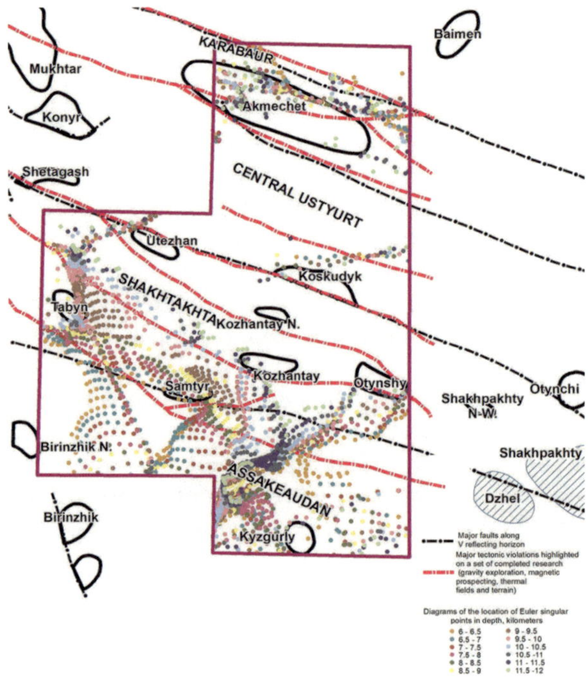

Figure 5.
Scheme of distribution by depth and density of Euler points (in interval 6–11 km) of the geomagnetic field.

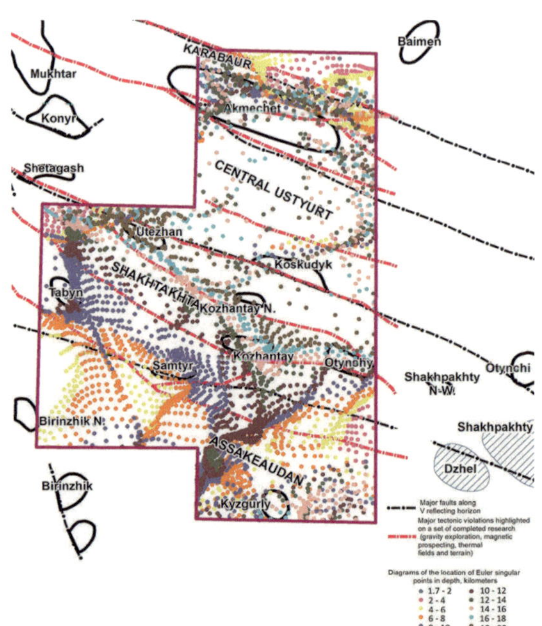

Figure 6.
Scheme of distribution by depth and density of Euler points (up to 20 km) of the geomagnetic field.

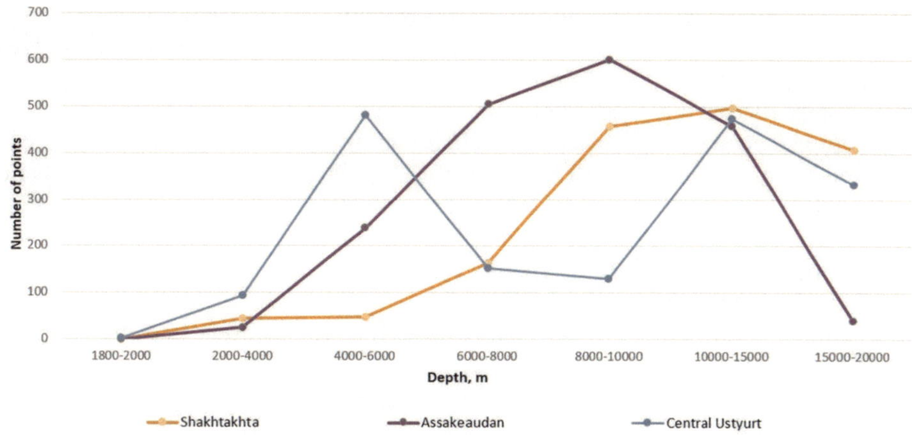

Figure 7.
Graph of the distribution of Euler points by regions in geomagnetic models.

6–11 km (**Figure 5**). The orientation of the zone of increased density of Euler points is consistent with the direction of deep faults, which is the basis for talking about their paragenetic relationship.

Additionally, the Central Ustyurt system of dislocations revealed the presence of another depth interval (10–15 km) with the highest concentration of Eulerian singular points (**Figure 7**). This finding indirectly suggests that the Karabaur swell has subsided in a northern direction under the North Ustyurt massif, resulting in a "double" crustal effect. However, this assumption needs further serious verification with the construction of three-dimensional geological and geophysical models.

Another feature of the specific Euler points has been revealed in all three regional tectonic structures considered above. A sharp or "jump-like" increase in the density of these points in the depth interval of 6–11 km is observed (**Figure 5**).

Considering the distribution of Euler points at various depths in geomagnetic models, it is possible to preliminarily conclude that the top edge of magnetically disturbing masses within the Central Ustyurt system of dislocations occurs at intervals of 4–6 km and 10–15 km depths, while the Assakeaudan Depression sees such occurrences at a depth of 8–10 km, and the Shakhpakhty tectonic step sees it at a depth of 10–15 км.

Analyzing the distribution density of Euler points in the depth interval up to 6 km (**Figure 4**), it is important to note their minimum values throughout the study region, which form small fields with an implicit orientation. It can be assumed that intrusions of the basic composition into the quasi-platform cover took place here.

And finally, the depth interval is up to 20 km. In the Assakeaudan trough, in the Shakhpakhty step, and in the Central Ustrta system of dislocations, there are no differences in the distribution density and orientations of the Euler points, which in turn indicates the basement of these tectonic elements (**Figure 6**).

2.9 Distribution of Euler points based on gravimetry data

The geological informativeness of the models built on the basis of the data of magnetically disturbing masses would be insufficient without the involvement of data about the depths of occurrence of gravity-disturbing objects.

The special Euler points, calculated in the Geosoft Oasis Montaj software using the algorithm of three-dimensional deconvolution [8], are concentrated near the edges of the anomalies, and correspond to the position and depth of anomaly-forming or gravitationally-disturbing bodies [9].

The regional structural elements are reliably identified with the distribution of density inhomogeneities in depth based on the distribution of Euler points, as well as with the involvement of other transformants of the gravitational field. Theoretical calculations show that even if the Euler points do not form dense clusters near anomalous bodies, at least in their vicinity, the distribution density of these points becomes the highest.

The scheme (**Figure 8**) and the table (**Table 2**) of the location of Euler singular points in the depth range from 0.6 to 8.0 km, as well as the graph of the distribution of these points in the vertical geological section (**Figure 9**) convincingly testify in favor of the density differentiation of rocks according to their depth.

The geological orientation of gravity-disturbing objects at the South Ustyurt is drastically differentiated by area. For example, in the 2–8 km depth interval in the northern part of the Shakhpakhty step, Euler points are grouped into bands of north-eastern geological strike (**Figure 9**).

In the Central Ustyurt dislocation system, special Euler points are concentrated in swathes of predominantly northwestern geological orientation.

In the south of the Shakhpakhty tectonic step and in the Assakaudan depression, these points form a cloud extending in a submeridional direction.

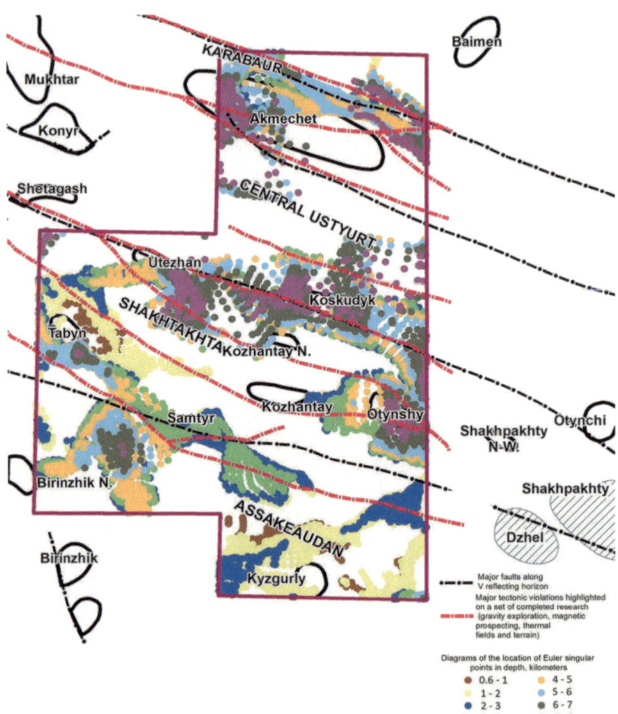

Figure 8.
Schematic diagram of the Euler points location in the gravitational field at depths up to 8.0 km (with an interval of 1.0 km). Legend: Black dotted lines represent faults complicating the structure of the V reflecting horizon. Dotted pink lines indicate deep faults identified by a set of geophysical data (gravity survey, magnetic survey, and thermal fields).

Regional structures	600–1000	Up to 2000 m	Up to 3000	Up to 4000	Up to 5000	Up to 6000	Up to 7000	Up to 8000
Central Ustyurt uplift	0	40	155	420	1765	810	255	240
Shakhpakhty fault terrace	0	42	245	460	500	585	610	280
Assakeaudan trough	190	1930	1390	1010	690	675	295	7

Table 2.
Locations of Euler points in the depth interval of 600–8000 meters.

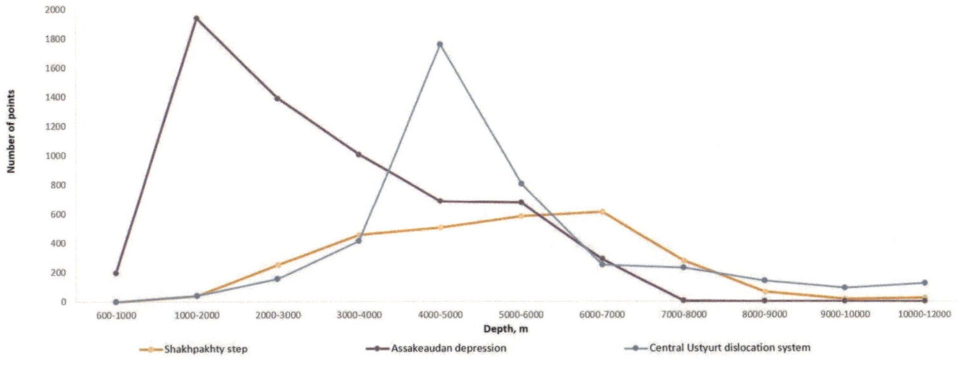

Figure 9.
Graph of the distribution of special Euler points based on gravimetry data (calculated in the Geosoft Oasis Montaj software).

At the Assakeaudan depression, the maximum number of Euler points on the surface of gravity-disturbing bodies is concentrated at depths up to 2000 m, in the Central Ustyurt dislocation system at depths up to 5000 m, and in the Shakhpakhty step at depths up to 6000 m (**Table 2**).

Figure 9 generally demonstrates a similar pattern, revealing significant variations in the depths of occurrence of gravity-disturbing bodies in all three regional structures, even at depths of 6000–7000 m. Deeper, these differences are significantly leveled. It can be assumed that this phenomenon is associated with a general transition from the formations of the quasi-platform cover to the rocks of the consolidated basement.

Considering the correlation of the depths of occurrence of gravitational and magnetically disturbing masses [10], it can be noted that accumulations of magnetically disturbing bodies in the Central Ustyurt dislocation system are observed in the depth intervals of 4000–6000 m. It is important to note that the depths of the occurrence of gravitationally and magnetically disturbing bodies coincide.

Based on CDP-2D seismic data, the depth of the basement surface at the Central Ustyurt dislocation system varies significantly, ranging from 4 to 8 km. The structures identified on this surface are linearly elongated, have a northwestern orientation, and are characterized by a general deepening in a southerly direction. In the area of the Akmechet local structure, the basement surface deepens to 9 km [11].

Consequently, the variations in the depths of the basement according to the CDP-2D seismic data are correlated in depth with the distribution of Euler points according to the gravity data.

It is important to note that in the Assakudan depression a serious discrepancy between the depths of occurrence of gravity-disturbing objects and magnetically disturbed masses was revealed.

The upper edge of the gravity-disturbing masses embeds at depths of 1000–2000 m and, apparently, reflects the transition from unconsolidated and weakly consolidated Lower Cretaceous rocks to more consolidated and epigenetically altered Upper Jurassic deposits. A similar phenomenon was revealed in some areas of the North Ustyurt region [7].

The upper edge of the magnetically disturbing masses here deepens to 8–10 km, which corresponds to the basement surface composed of rocks of basic and ultrabasic composition.

Interpretation data of the geomagnetic field anomalies are well correlated with the results of interpretation of the seismic CDP-2D data, which indicate that on the northern board of the Assakeaudan depression, the top of basement is deepened in a southerly direction from 7 to 10 km (at the Birinzhik local structure).

In contrast, a trend of uplift in the south-west direction can be observed along the top of the Upper Jurassic sediments (III reflecting horizon) from Kazgurly local structure, where it ranges from 2.6–2.8 km, to Birinzhik and Northern Birinzhik local structures, where it ranges from 1.7–2.0 km [11].

Therefore, we can talk about the correlation of the upper edge of the gravity-disturbing bodies with the top of Upper Jurassic sediments, while the upper edge of the magnetically disturbing masses is distinguished in the interval of depths of the top of basement.

At the Shakhpakhty step, a discrepancy was revealed in the position of the upper edges of the magnetic- and gravity-disturbing masses, which are submerged, respectively, to depths of 8–12 km and 6–7 km. The distribution pattern of these features suggests that there are differences in the depths of the basement and the quasi-platform cover.

This conclusion is confirmed by CDP-2D seismic data, based on which the depth of subsidence of the top of the Upper Paleozoic carbonate-terrigenous (quasi-platform cover) deposits varies in a wide range (from 3.8 km on the Tabyn structure, 4.5 km on the Samtyr and Utezhan structures, and up to more than 5.5 km on the Kozhantai and Otynshi structures).

The structures at the top of the Upper Paleozoic carbonate-terrigenous sediments here have polygonal shapes and mosaic character, as well as a general deepening trend in the southeast direction. The exception is the local structure of Tabyn with a northwestern orientation and brachianticlinal forms.

It is difficult to speak about the depth of the top of basement according to CDP seismic data, since this boundary does not have good acoustic rigidity.

Consequently, at the Shakhpakhty step, the upper edges of the gravity- and magnetically disturbing masses are characterized by maximum depth. At the Assakeaudan depression, these physical inhomogeneities have a large spread in depth, while in the Central Ustyurt system of dislocations, they are distinguished by small fluctuations in the depth of occurrence.

Conclusions obtained as a result of integrated interpretation of magnetometry and gravimetry data are confirmed by airborne gamma ray spectrometer (radiometric) data processed and interpreted by the thorium normalized method [12, 13].

Thus, the Tabyn, Kozhantai, Northern Kozhantai, Utezhan, and Kyzgyrly local structures are recognized as potentially promising for the search and exploration of

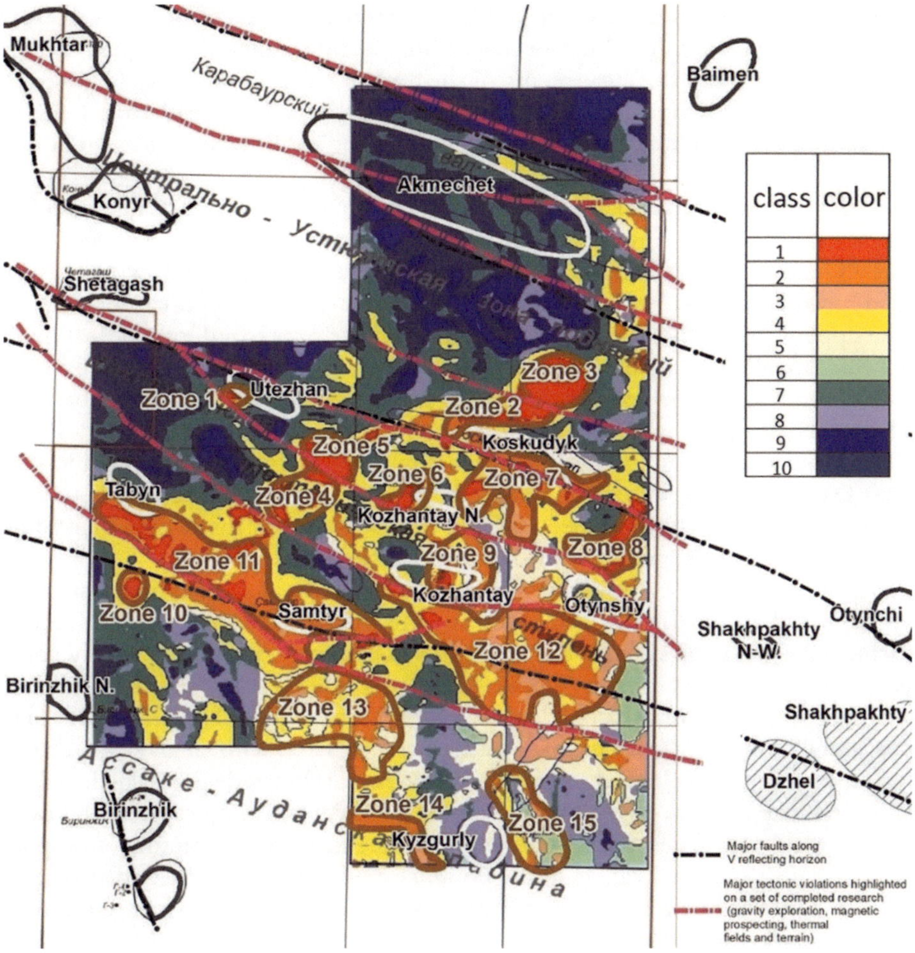

Figure 10.
Zonation of the area according to the RAE (radioactive) parameters (according to the algorithm of A.V. Petrov).

hydrocarbon deposits based on the data of the interpretation of the anomalous magnetic field and the parameters of potassium and uranium content (**Figure 10**). And further, it is recommended to carry out detailed study of these structures using the seismic CDP-3D and deep drilling.

3. Discussion

All of the above allows us to draw the following conclusions: According to the nature of the distribution of geomagnetic field transformants (using data on gravitational, radioactive, and geothermal potential fields), the Shakhpakhty tectonic step, the Assakudan depression, and the Central Ustyurt system of dislocations are distinguished.

3.1 Assakeaudan depression

It is identified by large positive anomalies in the northwest-orientated geomagnetic field, as well as high values in the transformants of magnetic anisotropy and the derivative of the angle of inclination of the geomagnetic field vector. Conversely, the minimum values are observed in the transformants of the analytical signal, the vertical derivative, and the autotracing of the axes of the ΔTa anomaly (**Table 1**).

The author's materials on study of gravitational, thermal, and radiation fields were involved in order to increase the reliability and geological information content of the research results.

Within the Assakeaudan depression, minimum or reduced values of the modules of the horizontal and vertical gradients of the Bouguer gravity anomalies, local negative gravity anomalies (recalculated to the upper half-space at a height of 2.5 km) are distinguished based on the results of a quantitative interpretation of the gravity field [9].

The upper edge of the causative magnetic masses drops to a depth of 8–10 km, which corresponds to the depths of the basement rocks (according to seismic CDP-2D) and reflects the level of maximum propagation of basic and ultrabasic rocks into the basement [14].

The spatial location of special Euler points in the gravity field calculated in the Geosoft Oasis Moptaj software [8] indicates the presence of rock density heterogeneity up to a depth of 8.0 km. Deeper, the differences in the depths of the Euler points in terms of density inhomogeneities are significantly leveled.

The upper edge of gravity-disturbing masses at the Assakeaudan depression stands out at depths of 1000–2000 m and is confined to the III reflecting horizon (according to seismic CDP-2D) and reflects the transition of unconsolidated and weakly consolidated Lower Cretaceous rocks to Upper Jurassic sediments [11].

Within the Assakeaudan trough based on the interpretation of geothermal fields [15] to a depth of 5 km, a vast area of low geothermal anomalies with small local maxima in the upper part of the geological section to a depth of 2.0 km is distinguished.

The evidence in favor of the completion of this depression by rocks with low thermophysical properties and a relatively isotropic geological structure is supported by the following: minimal values of the Bouguer anomaly horizontal and vertical gradient module transforms, local negative gravity anomalies (recalculated in the upper half-space at a height of 2.5 km), autotrace of the axes of the ΔTa magnetic field anomalies, and reduced values of TDR transformant [4, 14].

At the Assakeaudan depression according to the airborne gamma spectrometry survey data identified 4 anomalous zones with a relatively reduced background of total radioactivity, low isoconcentrations of radioactive potassium isotope; reduced uranium content in relation to the background values (**Table 1**). The formation of these zones is associated with the effect of hydrocarbon microseepage [16, 17] along the regmatic faults network and macrofracture systems, which indirectly indicate the presence of gas or oil fields [12, 18].

3.2 Shakhpakhty tectonic step

It is distinguished by low values of the transformants of the analytical signal, autotracing of the anomaly axes ΔTa, magnetic anisotropy, minimal values of the TDR transformant, high values of the intensity of negative anomalies ΔTa. Here, increased values of the transformants of the horizontal and vertical derivatives of the geomagnetic field are observed [5].

In general, these characteristics testify to the continuity of tracing the rock assemblages along their strike.

At the same time, in contrast to the Assakeaudan depression, at the Shakhpakhty step, increased values of the vertical gradient of gravity anomalies were revealed (**Table 1**), which indirectly may indicate an increased vertical variability of rocks.

Other characteristics of geophysical potential fields can be attributed to weakly and moderately intense local positive and negative Bouguer anomalies (recalculated to a height of 2.5 km) and increased values of intensity of negative anomalies, such as ΔTa (**Table 1**).

Within the Shakhpakhty step, the upper edge of the magnetically disturbing masses is submerged to depths of up to 8–12 km, while the gravity-disturbing bodies are deepened to 6.0–7.0 km. Apparently, the difference in their distribution demonstrates the difference in the occurrence depth of the tops of basement and the quasi-platform cover [11, 14].

Consequently, we can say that at the Shakhpakhty step, the upper edges of the gravity- and magnetically disturbing masses are distinguished by the maximum depth at the South Ustyurt region.

In turn, this fact is evidence in favor of the high total thickness of the sedimentary cover and the quasi-platform cover on the Shakhpakhty step, which unambiguously puts it in the category of prospective for hydrocarbon accumulation discoveries.

Shakhpakhty step is characterized by relatively large positive anomalies in the thermic fields, which indicates the predominance of rocks with high values of thermophysical properties (**Table 1**).

On the geological section that intersects the Shakhpakhty Step, where the gas field in the Uzbek part of Ustyurt has been explored, a relatively strong negative geothermic anomaly can be observed up to depths of 3000 m. This negative anomaly is further complicated in the upper part by two positive anomalies that reach maximum depths of 2200–2500 m.

However, it is known from literary sources [19, 20] that large gas field is marked by relatively negative thermal field anomalies, and positive anomalies in the upper part of the section, probably, are associated with compaction, providing a good "seal." However, this interpretation of thermic anomalies requires additional study in the process of integrated analysis, including data of high-precision gravity survey, seismic survey, and drilling.

On this step, according to airborne gamma spectrometry survey on the ground surface, the maximum number of zones (9 out of 15 in the South and Central Ustyurt regions) with anomalously low of radioactive potassium isotope and uranium concentration in relation to the background values was detected, which may indicate increased prospects for oil and gas content in local structures [12].

3.3 Central Ustyurt system of dislocations

Within this large tectonic element, the maximum values of the intensity of positive anomalies ΔTa, the transformants of the vertical derivative, and TDR vector inclination angle of the magnetic field, autotracing of the axes of the anomalies of this field, as well as increased values of the transformants of the analytical signal, magnetic anisotropy, and the horizontal derivative of the anomalies ΔTa were recorded [14].

In the surveyed area, there is a uniform characteristic of increased values of local gravity anomalies (recalculated in the upper half-space at a height of 2.5 km). Here, intermediate values of the transformants of the modules of the horizontal and vertical gradients of the Bouguer gravity anomalies are observed (**Table 1**).

Generally, the values of the transformants of the geomagnetic and gravity fields [9] may indicate an increased lateral and vertical heterogeneity of the rocks assemblage forming the Central Ustyurt dislocation system.

The maximum number of Euler points on the surface of gravity-disturbing bodies is concentrated at depths of −4000-5000 m.

The Central Ustyurt dislocation system displays gravity- and magnetically disturbing masses that coincide in depth and are confined to the basement surface, as identified by seismic CDP-2D [11].

An airborne gamma spectrometric survey conducted in this geostructure identified two zones with a relatively low background of total radioactivity. These zones exhibited low isoconcentrations of radioactive potassium isotope and reduced uranium content when compared to the background [12].

Thus, based on the foregoing, it can be argued that the integration of the magnetic field transformants with the data of aero gamma spectrometry and gravimetry indicates favorable prospects for the oil and gas potential of the Shakhpakhty step.

An indirect factor that testifies in favor of this is the large depth of immersion of the magnetically active layer associated with the basement rocks [10].

There was taken an attempt to tie-up the depth and character of distribution of magnetically active layers with areas potentially prospective for HC (hydrocarbon) accumulations to be detected.

In fact, the scientific novelty and practical significance of the obtained research results are the use of magnetic survey data for prospecting and exploration of hydrocarbon fields at the South Ustyurt.

Thus, the transformants of the initial geomagnetic field presumably increase reliability in detection of anomalous objects, and may be considered as an extra exploration criterion in prospecting and exploration of HC fields.

4. Recommendations for exploration

The focus of attention in the regional study should be placed on conducting exploration on local structures at the Shakhpakhty step, within which there is a deep plunge of the basement, the presence of thick strata of sedimentary cover and quasi-platform cover, and the increased positive anomalies of the thermal field were identified [21].

The integrated interpretation of data on the thermic field and the transformants of the gravity and magnetic fields, as well as aero gamma spectrometry (processed and interpreted by the thorium normalization method), suggest promising oil and gas prospects in local structures such as Utezhan, Kozhantai, Northern Kozhantai, and Otynshi. This is attributed to favorable historical-geological, structural, and lithofacies factors (**Table 3**).

The location of these local structures is given in the figures above in the text.

The CDP-3D seismic survey and deep exploratory drilling with obligatory penetration of sediments of the quasi-platform cover are recommended on these local structures in order to explore new hydrocarbons accumulations.

5. Conclusion

The depth and nature of the spread of the magnetoactive layer in areas prospective for the detection of HC accumulations were successfully related using various the

Potential fields	Transforms	Structures of Jurassic-Cretaceous deposits revealed by CDP-2D seismic survey			
		Utezhan	Kozhantai N.	Kozhantai	Otynshi
Geomagnetic anomaly field	Anomalous magnetic field reduced-to-the-pole, nT.	From −5.50 to 2.21	From −5.04 to 1.93	From 21.41 to 53.67	From 8.18 to 31.04
	Analytic signal, nT/m.	From 0.006 to 0.008	From 0.005 to 0.008	From 0.005 to 0.008	From 0.005 to 0.012
	Tilt derivative (TDR) of the magnetic field anomalies, radian.	From −1.56 to −1.23	From −1.36 to −0.94	From −0.80 to −0.05	From −1.36 to −0.24
	Vertical derivative of the anomalous magnetic field, nT/m.	From −0.004 to −0.001	From −0.003 to −0.001	From −0.003 to 0.00001	From −0.004 to 0.001
	Horizontal derivative of the modulus magnetic field, nT/m.	From 0.000 to 0.002	From 0.001 to 0.003	From 0.005 to 0.006	From 0.003 to 0.005
	Transformation of the magnetic field anisotropy, cu.	From 0 to 0.0003	From 0 to 0.0001	From 0.0005 to 0.001	From 0 to 0.0005
	Tracing of axes of magnetic field anomaly, cu.	From −0.57 to 0.53	From −0.68 to 0.26	From −1.92 to 0.39	From −1.77 to 0.34
Gravity anomaly field	Module of horizontal derivate of Bouguer anomaly, mGl/m.	From 0.0002 to 0.0010	From 0.0003 to 0.0008	From 0.00003 to 0.0007	From 0.0004 to 0.0010
	Module of the vertical derivate of Bouguer anomaly, mGl/m.	From −0.0004 to 0.0005	From −0.0007 to 0.0001	From −0.0006 to 0.0004	From −0.0004 to 0.0008
	Local gravity anomaly (recalculated in the upper half-space at a height of 2.5 km), mGl.	From −0.84 to −0.23	From −1.19 to −0.83	From −1.15 to −0.70	From −1.03 to −0.02
	Anisotropy transform, the accent of maximum gravity anomaly. Sliding window size 5 km x 2 km, cu.	From 0.1 to 0.12	From 0.1 to 0.11	From 0.1 to 0.14	From 0.1 to 0.14
Thermal anomaly field	Thermal field according to satellite imagery.	From 6266.32 to 6428.80	From 6281.45 to 6404.68	From 6333.09 to 6509.25	From 6442.89 to 6556.22
	Map of the regional component of the thermal field.	From 6311.40 to 6357.90	From 6369.59 to 6399.50	From 6380.17 to 6449.76	From 6495.36 to 6524.38
	Map of the local component of the thermal field.	From −26.54 to 50.10	From −76.65 to −17.70	From −38.34 to 51.75	From −49.79 to 29.48

Potential fields	Transforms	Structures of Jurassic-Cretaceous deposits revealed by CDP-2D seismic survey			
		Utezhan	Kozhantai N.	Kozhantai	Otynshi
	Map of the thermal field recalculated in the lower half-space. A cut at 1500 m.	From −3.1 to 10.6	From −5.3 to 2.3	From −3.6 to 4.2	From −7.3 to 3.6
	Map of the thermal field recalculated in the lower half-space. A cut at 2500 m.	From −0.4 to 10.4	From −7.3 to −0.6	From −4.4 to 4.5	From −7.6 to 3.8
Radiogeochemical anomalies field	Potassium value (K), %	From 0.57 to 1.05	From 0.71 to 1.04	From 0.68 to 1.07	From 0.70 to 1.12
	Thorium value (Th), ·10–4%	From 1.54 to 3.30	From 1.94 to 3.14	From 1.92 to 3.29	From 1.89 to 3.45
	Uranium value (U), ·10–4%	From 1.70 to 2.56	From 1.64 to 2.46	From 1.61 to 2.47	From 1.90 to 2.51
	Dose rate gamma radioactivity (DR)	From 19.27 to 31.88	From 22.90 to 31.64	From 21.30 to 32.15	From 21.89 to 32.91
	U/K, cu	From 1.90 to 3.15	From 2.01 to 2.83	From 2.02 to 3.04	From 1.99 to 3.11
	K/Th, cu	From 2.66 to 3.18	From 2.70 to 3.12	From 2.68 to 3.09	From 2.58 to 3.13
	U/Th, cu	From 0.88 to 1.78	From 1.04 to 1.63	From 0.93 to 1.62	From 0.88 to 1.67

Note: TF—thermic field; MF—magnetic field.

Table 3.
Statistics on transform values for anomalous magnetic, gravity, thermal, and radiogeochemical field transforms by local structures over the study region.

transformants of this field, such as analytical signal magnetic anisotropy, autotracing axes of magnetic field anomalies, increased values of negative anomaly ΔTa, and horizontal and vertical magnetic field gradients, as well as the minimum values of the TDR.

The results of an integrated analysis of the geomagnetic, gravitational, thermal, and radiochemical fields allowed the authors of this chapter to study the behavior and characteristics of tectonic faults, and to make some judgments about the depth of occurrence of gravity and magnetically disturbing masses, about the degree of geological heterogeneity of large geostructures, and to trace the nature of the manifestation of local structures in geophysical fields.

The Shakhpakhty step with a relatively increased thermal field in some local structures is quite clearly distinguished into the regional gravity, geomagnetic, and thermal fields, and has a continuation in the Uzbek part of this step.

The Shakhpakhty step is bounded by deep faults, identified by the gradients of anomalies of the regional geophysical fields. Three major regional faults are traced in the northwestern direction. On the southeastern flank of these faults, the Shakhpakhty and Dzhel gas fields are explored within the Jurassic sediments. On the western flank, the local structures Otynshi, Kozhantai, Northern Kozhantai, Utezhan, and others close to them in geological structure are isolated.

These local structures can be considered as prospective in terms of the localization of HC accumulations, if there is a good seal of rocks in the upper part of the predicted reservoirs. All of the aforementioned criteria give reason to recommend local structures within the Shakhpakhty step for priority study.

Central Ustyurt system of dislocations is characterized by elevated deformation of sedimentary rock assemblage that does not contribute to the preservation of hydrocarbon accumulations.

The Assakeaudan depression is characterized by thick sedimentary rock assemblages, which generated mostly by the oil-source suites. Faults that disrupt the integrity of the geological formations of this depression are considered as conduits for migration of hydrocarbons, deep fluids, heat, and mass transfer.

This statement is confirmed by aero gamma spectrometry data, processed and interpreted by the method of thorium normalization. A total of 15 anomalous zones were identified in the study region, 9 of which are located in the Shakhpakhty step, 2 in the Central Ustyurt system of dislocations, and 4 in the Assakeaudan depression.

The selected anomalous zones have low values of isoconcentrations of radioactive potassium isotope and reduced uranium concentration in relation to the background values and are considered as indicative criteria for the possible presence of hydrocarbon accumulations.

Thus, our studies of the transformants of the magnetic, gravitational, and geothermal fields, airborne gamma spectrometry data (processed and interpreted using the thorium normalization method) unequivocally testify in favor of the prospects for the oil and gas potential of the local structures Utezhan, Kozhantai, Northern Kozhantai, Otynshy, taking into account favorable historical and geological, structural, and lithofacies factors.

The author's study on advanced (relatively cheap) geophysical methods will provide a dependable foundation for enhancing and specifying the geological and structural-tectonic models of South Ustyurt. This will involve using existing information on the spatial distribution of already known oil and gas fields as well as serving as a basis for designing geological exploration to prospect and explore new hydrocarbon accumulations, using expensive and "heavy" methods of seismic exploration CDP-3D and deep drilling.

Author details

Abetov Auez Egemberdievich[1]*,
Yessirkepova Sharbanu Bakhytovna[1] and Julia Barbosa Curto Ma[2]

1 Kazakh National Research Technical University named after K. I. Satpayeva, Almaty, Kazakhstan

2 University of Brasilia, Brasilia, Brazil

*Address all correspondence to: abetov.auez@mail.ru

References

[1] Daukeyev SZ. Deep structure and mineral resources of Kazakhstan. In: Volume III Oil and Gas. Almaty; 2002. p. 248

[2] Ozdoev SM. Prospects of oil and gas content in sedimentary basins of Kazakhstan. NEWS OF NAS RK, Series of geology and technical sciences. 2012;**1**:61-76

[3] Blakely RJ. Potential Theory in Gravity and Magnetic Applications. Cambridge: Cambridge Univ. Press; 1996. DOI: 10.1017/CBO9780511549816

[4] Castro FR, Oliveira SP, de Souza J, Ferreira FJF. Combining tilt derivative filters: New approaches to enhance magnetic anomalies. Brazilian Journal of Geophysics. 2018;**37**. DOI: 10.22564/rbgf.v36i3.1956

[5] Saada SA. Edge detection and depth estimation of Galala El Bahariya plateau, Eastern Desert-Egypt, from aeromagnetic data. Geomech. Geophys. Geo-energ. Geo-resour. 2015;**2**:25-41. DOI: 10.1007/s40948-015-0019-6

[6] Petrov AV, Nikitin AA. Coscad 3Dt. Complex of Spectral-Correlation Analysis of Geophysical Evidence, user's Manual Part I. Moscow: Russian State Geology Prospecting University; 2003

[7] Abetov A, Niyazova A, Saurikov Z. 3D modeling of Euler's points for Geodensity and geomagnetic models of north Ustyurt region in Geosoft Oasis Montaj software. NEWS OF NAS RK, Series of geology and technical sciences. 2017;**6**:171-177

[8] Montaj MAGMAP Filtering 2D Frequency Domain Processing of Potential Field Data Extension for Oasis Montaj. Vol. 7.12013

[9] Matusevich AV. Structure of the local gravity field as a basis for zoning salt domes Caspian depression. News of NAS RK. Series of geology and technology sciences. 2006;**5**:65-76

[10] Serkerov SA. Gravity and Magnetic Prospecting in Oil and Gas. Moscow: Oil and Gas. Gubkin Russian State University of Oil and Gas; 2006. pp. 501-512

[11] Abetov AE, Yessirkepova Sh B, Curto MJ. Gravity Field Transforms at the Exploration for Hydrocarbon Field in the Southern Part of the Ustyurt Region. Vol. 3. NEWS OF NAS RK, Series of geology and technical sciences; 2022. pp. 17-31

[12] Abetov AY, Essirkepova SB, Kozhamsugirov D. Effectiveness of aerogamma-spectrometric research in solving applied problems of oil and gas geology. Geologiya i okhrana nedr. 2021, 2021;**78**(1):55-62

[13] Saunders DF, Burson KR, Branch JF, Thompson CK. Relation of Thorium-Normalized Surface and Aerial Radiometric Data to Subsurface Petroleum Accumulations. Vol. 58, N. 10. Tulsa, OK, United States: Society of Exploration Geophysicists; 1993. pp. 1417-1427

[14] Abetov AE, Yessirkepova Sh B, Curto MJ. Geomagnetic Field Transforms and their Interpretation at Exploration for Hydrocarbon Field in the Southern Part of the Ustyurt Region. Vol. 6. NEWS OF NAS RK, Series of geology and technical sciences; 2021. pp. 6-14

[15] Kronberg P. Remote study of the earth. In: Fundamentals and Methods of Distance Research in Geology. Moscow: World. Trans.; 1988. p. 343

[16] Saunders DF, Burson K, Thompson CK. Model for hydrocarbon microseepage and related near-surface alterations. AAPG Bulletin. 1999;**83**(1): 170-185

[17] Curto JB. The Role of Aero Geophysics for Detecting Hydrocarbon Microseepages and Related Structural Features: The Case of Remanso Do Fogo. Vol. 77, N. 2. Brazil: Geophysics; 2012. pp. B35-B41

[18] Klusman RW. Soil Gas and Related Methods for Natural Resource Exploration. New York: John Wiley; 1993. 483 p

[19] Shilin BV. Thermal Surveying in the Study of Natural Resources. Moscow: Gidrometeoizdat; 1980

[20] Gornyi VI, Shilin BV, Yasinskii GI. Thermal Aerospace Survey. Moscow: Nedra; 1993 128 p

[21] Abetov AE, Yessirkepova Sh B, Curto MJ. Remote sensing at the study of the thermal field of the South Ustyurt Region to find hydrocarbon deposits. NEWS OF NAS RK, Series of Geology and Technical Sciences. 2023;**2**:6-16

Section 6

LiDAR Applications

Chapter 9

Spaceborne LiDAR Surveying and Mapping

Fang Yong, Zhang Li, Gong Hui, Cao Bincai, Gao Li and Hu Haiyan

Abstract

Laser point cloud data have the characteristics of high elevation accuracy, fast processing efficiency, strong three-dimensional (3D) vision, and wide application fields. It will be one of the core datasets of the new generation national global topographic database. The rapid advancement of spaceborne laser earth observation technology allows the collection of global 3D point cloud data, which has brought a new breakthrough in the field of satellite-based earth observation, and its significant advantages of all-day time, high accuracy and high efficiency will lead the future development of space precise mapping technology. This chapter firstly introduces the principle and development status of satellite-based LiDAR technology, then presents the basic technical framework of satellite-based LiDAR 3D mapping, and analyzes the data processing methods of spaceborne photon point clouds, and finally, focuses on the application research in various fields including precise geolocation of combined with satellite images, fusion of multi-source topographic information, polar mapping, 3D objects reconstruction, and shallow sea topographic mapping, etc.

Keywords: global laser point cloud, global control point library, global 3D elevation data, 3D digital geospatial framework, global topographic database, 3D precise geometric positioning, multi-source topographic information fusion, polar mapping, 3D object reconstruction, shallow sea topographic mapping

1. Introduction

In recent years, LiDAR (Light Detection and Ranging) has been developing rapidly as a new generation of precise earth observation technology. The satellite-based LiDAR system uses satellite as the platform and photon-counting LiDAR as the main payload to detect global surface 3D information around the clock, precisely determine laser point positions in near real time, and simultaneously collect 3D point clouds (active SLAM) in the mission region, as shown in **Figure 1**, providing a new and efficient means to rapidly implement global 3D information mapping (including:

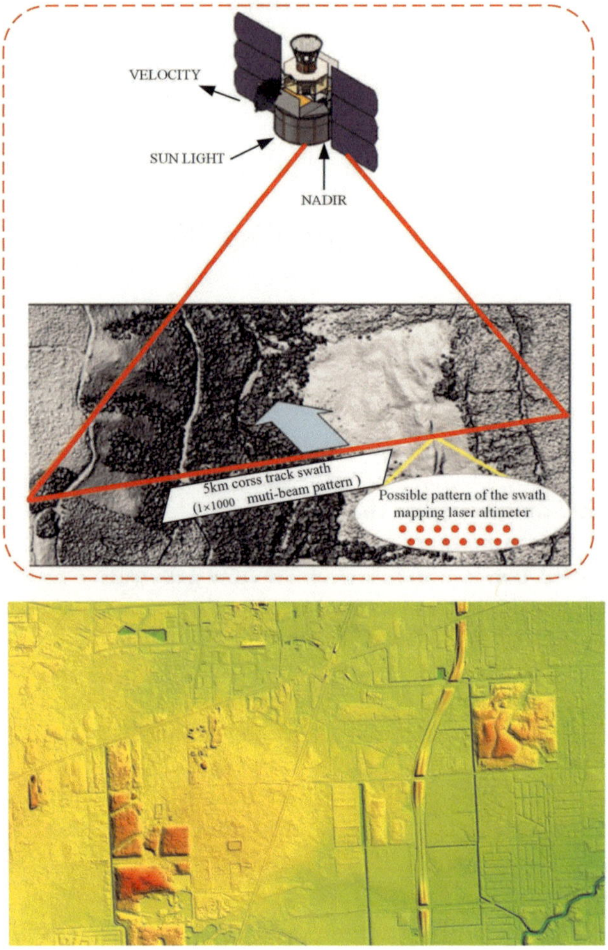

Figure 1.
Schematic diagram of satellite-based LiDAR measurement. (TOP: LIST satellite for example; Bottom: DSM).

high-precision laser control points, 3D digital surface model (DSM), and digital elevation model (DEM)).

Compared with the existing or developing optical or microwave remote sensing mapping satellites, it has the advantages of high accuracy of observation data and fast information acquisition efficiency. First, the elevation accuracy is improved by 5–10 times; second, the data processing is highly automated; and third, the overall acquisition cost is significantly reduced. The development of satellite-based LiDAR measurement means can effectively improve the overall geometric accuracy of earth observation, provide basic 3D topographic data support for the comprehensive application of various types of remote sensing satellite images; fill in the gaps of geospatial information in the polar regions; and provide high-precision 3D frame information support for precise location services [1–8].

The rapid development of satellite-based LiDAR measurement technology has brought a new breakthrough in the field of satellite-based earth observation, and the significant advantages of all-day, high-precision and high-efficiency will definitely lead the future development direction of aerospace remote sensing and mapping technology [5].

2. Principle and development status of satellite-based LiDAR technology

2.1 Basic principle

The satellite-based LiDAR measurement is performed by transmitting laser pulses at a certain frequency from the satellite-based laser to the ground, and the laser beam crosses the atmosphere and is scattered by the terrain surface, producing a weak backscattered echo, which is received by the telescope on the satellite-based LiDAR, and the distance value between the laser and the detection target is calculated through photoelectric signal conversion and time measurement, and then combined with the information of satellite attitude, platform position, and laser pointing to finally obtain precise three-dimensional spatial coordinates of the laser footprint point, as shown in **Figure 2**. The satellite-based LiDAR measurement belongs to the direct active acquisition of surface elevation information, which is different from the traditional indirect reconstruction measurement mode of remote sensing imaging mapping, reducing the time of post-processing process and improving the efficiency of surface 3D information acquisition.

2.2 Status of development

Satellite-based laser measurement equipment was first used in deep space exploration, such as Apollo-15, 16, 17 (1970, Moon), Clementine (1994, Moon), MGS (1996, Mars), LRO (2009, Moon), OSIRIS (2016, asteroids), Japan SELENE (2007, Moon), Chinese CE-1, 2, 3, 4 (since 2007, Moon) [9], mainly used for surface topography measurements and landing site selection. Due to the large differences between deep space exploration and earth observation in terms of detection environment and accuracy requirements [10, 11], this paper only focuses on earth observation LiDAR systems. As shown in **Table 1**, the existing and planned LiDAR systems for earth observation at worldwide [5, 12–16] mainly include SLA, ICESat, ICESat-2, ZY3–02, GF-7, terrestrial ecosystem carbon monitoring satellite, GEDI, and LIST.

NASA launched the first laser measurement satellite, the Ice, Cloud, and Land Elevation Satellite (ICESat-1), in January 2003, with the Geoscience Laser Altimeter System (GLAS) as the primary payload, with the main mission of measuring land

Figure 2.
Principal diagram of satellite-based LiDAR measurement.

Satellite/Payload	Time	Country	Detection mode	Beams	Pulse width (ns)	Sampling distance (m)	Size of footprint (m)	Elevation accuracy (m)	Applications
SLA-01/02	1996/97	US	Pulse	1	10	750	100	1.5	Global Elevation Control Points
ICESat/GLAS	2003	US	Pulse	1	6	170	70	0.15	Sea ice, atmosphere, land, vegetation, etc.
ZY3–02	2016	China	Pulse	1	7	3500	50	1.0	Experimental altimetry
ICESat-2/ ATLAS	2018	US	Photon counting	6	1.5	0.7	<17.5	0.1	Polar regions, ice sheets, atmosphere, land, vegetation, oceans, etc.
GEDI	2018	US	Pulse	8	14	60	25	1.0	Forest Biomass Monitoring
GF-7	2019	China	Pulse	2	4–8	2900	30	1.0	Generalized Elevation Control Point
Terrestrial Ecosystem Carbon Monitoring Satellite	2022	China	Pulse	5	7	200	25–30	1.0	Forestry carbon inventory monitoring, generalized elevation control points
LIST	To be launched	US	Photon counting	1000	1	0.7	5	0.1	Global Digital Elevation Model

Table 1.
Main technical specifications and applications of earth observation laser altimetry instruments at worldwide.

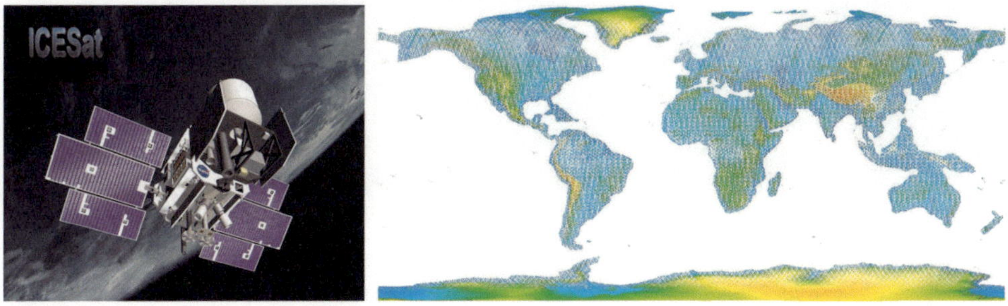

Figure 3.
ICESat and its global laser point cloud data (395 million points, 1.36 GB).

topography, cloud height and vertical structure, and polar ice caps [17]. During its 6.5 years of operation, the ICESat-1 laser altimeter has acquired a large amount of high-precision surface data of the earth, as shown in **Figure 3**, providing a valuable data source for scientific research and practical applications in many fields [18, 19]. Of particular note in the field of global mapping is the ability of surface elevation measurements to construct accurate medium-resolution digital elevation models of the Earth or to combine with optical stereo mapping/InSAR mapping satellite results to significantly improve the elevation accuracy of 3D terrain models [20]. Several studies have shown that the surface laser points acquired by ICESat-1 can be used as auxiliary control data [21, 22], and its elevation measurement accuracy reaches 0.1 m magnitude, which significantly improves the accuracy of aerotriangulation and mapping under uncontrolled conditions, especially in the elevation accuracy [23]. China has added similar laser altimeter payloads to the newly developed Gaofen 7 and Gaofen 14 high-precision stereo mapping satellites [24, 25], which provide support to ensure the accuracy of 1:10,000 scale elevation surveying under satellite conditions.

Given the outstanding performance of the ICESat-1 satellite, the follow-on ICESat-2 laser measurement satellite mission became one of the highest priority satellite observation missions recommended by the National Research Council for the period 2010–2020 [26]. ICESat-2 was successfully launched in September 2018 with a primary payload of the Advanced Topographic Laser Altimeter System (ATLAS), which adopts a photon-counting, high-frequency, micro-pulse, multi-beam laser measurement scheme with a 17 m footprint on the ground and a sampling interval of 0.7 m in the along-track direction. As shown in **Figure 4**, the accuracy and reliability of the earth observation results have been greatly improved compared with the 70 m footprint and 167 m along-track sampling interval of ICESAT-1 [27, 28]. In the 2 years of ICESAT-2 operation, the global surface data collection density is 70 cm apart in the along-track direction, and the maximum spacing is less than 2 km in the vertical track direction on the equator, which provides the possibility to construct a topographic elevation model with higher plane accuracy and resolution from regional to global scales.

The planned launches of laser earth observation satellites are the Terrestrial Ecosystem Carbon Monitoring Satellite (TECMS) and the U.S. LIST (LiDAR Surface Topography) program. The carbon monitoring satellite is the first satellite mainly serving forestry in China, carrying LiDAR and high-resolution multi-angle multispectral cameras to achieve forest height and biomass inversion through synoptic measurements; LIST is proposed to use photon counting detection system to obtain global topographic information of 5 m grid size and 10 cm elevation accuracy with 1000 beams, as well as surface elevation changes of forests, lakes, and ice caps [29].

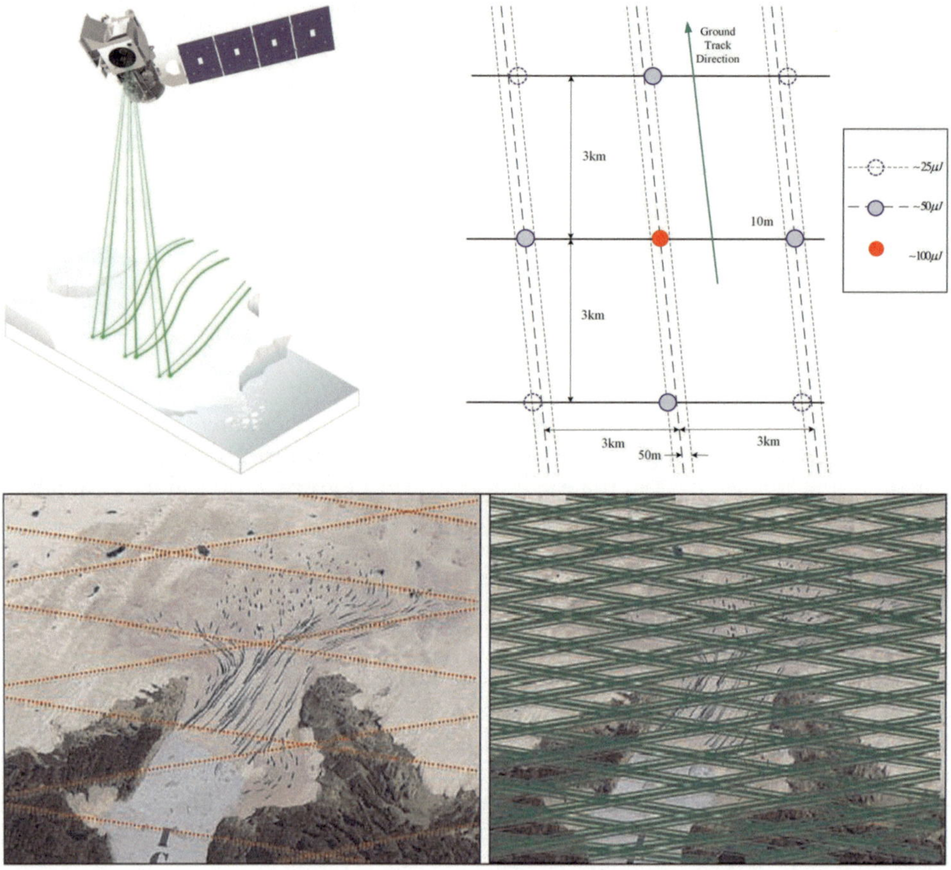

Figure 4.
ICESat-2 satellite, ground beam geometry, comparison of GLAS and ATLAS laser footprint distribution (1500 times higher sampling density).

The satellite-borne high-resolution Earth observation LiDAR is still a very cutting-edge engineering research direction internationally, and foreign research institutions, including the United States, have taken it as an important research content and development direction for future Earth observation. NASA's 20-year development plan for laser imaging radar satellites is shown in the **Figure 5**.

3. Basic technical framework of spaceborne LiDAR 3D mapping

The preferred approach to the acquisition of global laser point cloud data is to use spaceborne instruments. Since the beginning of the twentieth century, two generations of satellite-based LiDAR measurement technology have been developed [30, 31]: the first generation is represented by ICESAT-1, which uses a laser linear detection regime characterized by high-energy pulses, full waveform detection, and high accuracy of elevation measurement to the decimeter level [32]. Limited by the volume of power consumption, it is generally a single beam or several beams with a large laser footprint, which to some extent affects the planimetric positioning accuracy to about 10 m [33]; at the same time, the sampling frequency is low, generally 2–3 Hz, and the density of acquisition points is low, which requires long-period continuous

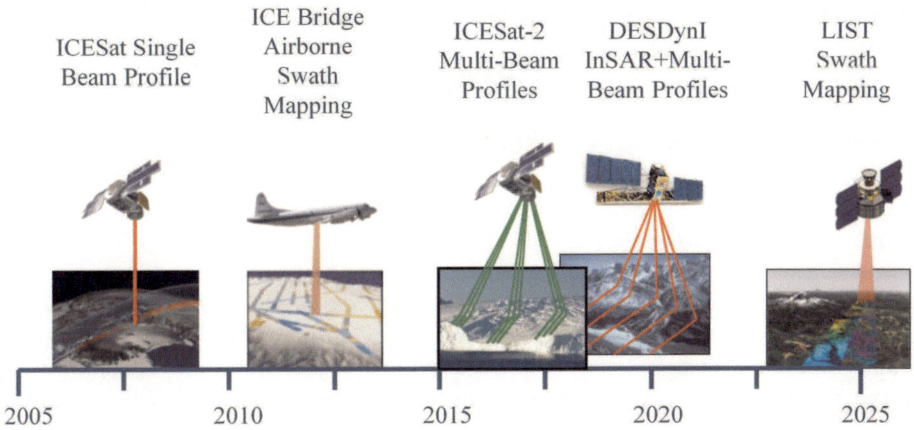

Figure 5.
NASA's 20-year development plan for laser imaging radar satellites.

observation to achieve dense global coverage. The second generation, represented by ICESAT-2, adopts a photon-counting laser detection system [34, 35], characterized by micro-pulse and multi-beam detection, which improves the detection effectiveness by three orders of magnitude while maintaining high accuracy of elevation measurement, and significantly increases the sampling frequency and data sampling density, providing an effective technical way to achieve global high-precision and high-density 3D point cloud data acquisition [28]. With the support of the National Major Science and Technology Project "High Resolution Earth Observation System," China has broken through the key technology of second-generation satellite-based LiDAR measurement after more than 10 years, which has laid a solid foundation for the development of independent laser mapping satellites.

Figure 6 shows the basic technology framework for rapid construction of 3D digital geospatial information based on global laser point cloud. The framework

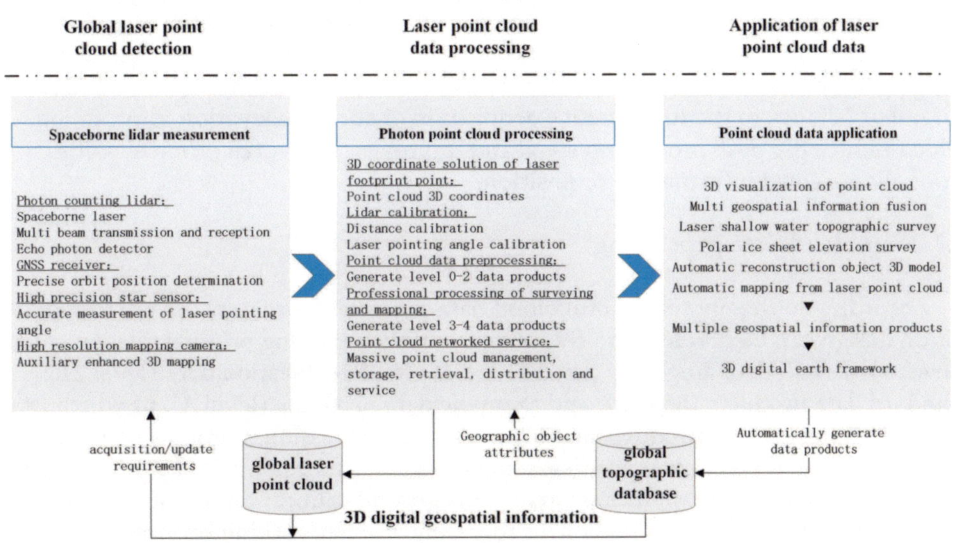

Figure 6.
The setup diagram of 3D digital geospatial information framework to rapid reconstruct based on global laser point cloud.

system mainly includes three parts: global photonic laser point cloud data acquisition, data processing and data application, and the ultimate goal is to form a global point cloud database to support the construction and update of global mapping database.

3.1 Global laser point cloud acquisition

The satellite-based LiDAR measurement is performed by transmitting laser pulses at a certain frequency from the satellite-based laser to the ground, and the laser beam crosses the atmosphere and is scattered by the ground surface, producing a weak backscattered echo, which is received by the telescope on the satellite-based LiDAR, and the distance value between the laser and the detected ground surface is calculated through photoelectric signal conversion and time measurement, and then combined with the satellite attitude, platform position, laser pointing, and other information to finally obtain precise 3D spatial coordinates of the laser footprint point. It mainly involves the key technologies such as long-life on-board laser, multi-beam laser transmission and reception, photon level signal detection, precision orbit position determination, laser pointing accuracy determination, and joint active-passive 3D measurement. The major key technologies involved are long-life on-board laser, multi-beam laser transmission and reception, photon level signal detection, precision orbital position determination, precise determination of laser pointing, and joint active-passive 3D measurement, etc. The technology is characterized by the requirement of laser pointing accuracy to subangular second level and distance measurement accuracy to centimeter level at working altitude of hundreds of kilometers and the realization of 3D coordinate measurement accuracy of laser footprint points on the plane to meter level and elevation to decimeter level. Meanwhile, in order to improve the efficiency and accuracy of global mapping, it adopts a new system of photon counting LiDAR, which reduces the laser footprint size from tens and dozens of meters to the meter level and increases the sampling frequency from a few Hz to 10,000 Hz, improving the detection effectiveness by thousands of times compared with the traditional linear detective system, providing support for the rapid acquisition of high-density and high-quality point clouds. Unlike the traditional optical or microwave remote sensing mapping through the indirect measurement mode by the imaging to achieve three-dimensional reconstruction, the satellite-based LiDAR measurement belongs to the direct active acquisition of surface elevation information, which reduce the post-processing steps and improves the overall efficiency of surface three-dimensional information acquisition.

3.2 Point cloud data processing

According to the photonic point cloud data processing process, the product definition design can be divided into five levels, mainly including original telemetry data (level 0), format decoding data (level 1), point cloud geolocation data (level 2), standard data products (level 3), and thematic data products (level 4), of which levels 0–2 are pre-processing products and levels 3–4 are professional processing products, involving the main technical processes including laser foot point 3D coordinate solution, LiDAR ranging and pointing parameters ground calibration, point cloud data pre-processing, mapping professional processing and networking services. Level 1 processing is to decode the raw package data, format conversion, and data cataloging to obtain the standard format data. The raw standard format data include photon time-of-flight data obtained after instrument delay and other corrections, laser emission position, and pointing data obtained using precision positional data and calibration data. Level 2 processing is to obtain the laser footprint longitude, latitude, and elevation values, and denoising and pre-classifying the point cloud [36], including (2A) un-noised point cloud and (2B) denoised point cloud. Level 3 processing is to obtain control point data, target area DSM, and polar DSM products after adjustment and gridding, including (3A) control point data, (3B) DSM, (3C) DEM, and (3D) object 3D model. Level 4 processing is for specific application needs, fusion of optical, microwave images and DSM, and other multi-source data to generate thematic products for specific applications, as well as extraction of polar ice cover, atmosphere, vegetation, lakes, and other information in point clouds, and other corresponding thematic products are obtained through specialized professional processing of mapping.

3.3 Point cloud data applications

As a direct digital representation of the global 3D physical world, laser point clouds have been applied in a wide range of directions [37–41], mainly including massive global laser point cloud 3D visualization, multiple terrain information fusion, shallow sea topography, polar elevation measurement, automatic 3D reconstruction of objects, and digital mapping based on point clouds, etc., to realize diverse spatial information product making and provide support for the construction of 3D digital earth framework. Efficient storage management retrieval and visualization of massive point cloud data are the most direct and effective applications. At present, there are successful solutions to support the storage management and visualization application of 3D point clouds with a global data volume of 100 petabytes, such as Bentley Pointools [42] and Euclideon udStream [43], whose 3D engine has the characteristics of loading unlimited spatial data in seconds to achieve rapid application. Secondly, point cloud control surveying realizes multi-source observation data fusion application. Cloud control photogrammetry has been realized in a number of system construction and engineering applications, point cloud support for multi-source terrain information, and three-dimensional model fusion will be the next important research direction, to provide an effective way to quickly establish a large range of consistent accuracy, more rich information-type three-dimensional geospatial information framework. In addition, point cloud mapping to achieve automatic acquisition of 3D information from laser point clouds and transformation into geographic entity representation with structure and function has become a major application direction. Dense point cloud and fully automatic processing can be applied to urban 3D modeling, coastal zone topographic survey, polar elevation mapping, road infrastructure maintenance monitoring, and forest resources survey, etc. It generates multifaceted geospatial information products including 3D models of buildings, digital ground models and digital surface models, forestry thematic products, etc., which can provide support for global mapping database construction and thematic element information update.

4. Data processing methods

4.1 Photon geolocation

The basic principle of satellite-based laser measurement and geometric positioning is that the laser beam is transmitted by the satellite and received by the satellite after reflection from the ground, and the time interval between the laser transmission and reception t is calculated. The propagation speed of light is c, and the one-way transmission distance of the laser is $\rho = ct/2$, and the three-dimensional coordinates of the laser footprint can be obtained by combining the satellite position and attitude information obtained from the GNSS positioning instrument and the star sensor on board the satellite [44]. The rigorous geometric model is shown in **Figure 7**, where P_{laser} is the reference point of laser emission, P_{GNSS} is the GNSS phase center, O_{Body} is the satellite center of mass, and P_{Ground} is the laser ground footprint (Bounce Point Location).

Laser footprints were precisely geolocated and point clouds were generated using on-orbit calibrated laser ranging parameters, laser pointing parameters, and precision attitude/orbit data, taking into account geophysical corrections such as atmospheric delays and tides. The rigorous geometric location equation for satellite-based LiDAR is as follows:

$$\begin{pmatrix} X_{spot} \\ Y_{spot} \\ Z_{spot} \end{pmatrix}_{ITRF} = \begin{pmatrix} X_s \\ Y_s \\ Z_s \end{pmatrix}_{ITRF} + R_{ICRF}^{ITRF} R_{BOD}^{ICRF} \left[\begin{pmatrix} \Delta X_{ref} \\ \Delta Y_{ref} \\ \Delta Z_{ref} \end{pmatrix} + \rho \begin{pmatrix} \sin(\theta + \Delta\theta_{1\_i}) \cos\alpha \\ \sin(\theta + \Delta\theta_{1\_i}) \sin\alpha \\ \cos(\theta + \Delta\theta_{1\_i}) \end{pmatrix} \right] \quad (1)$$

where $(X_s\ Y_s\ Z_s)_{ITRF}^T$ is the coordinate of the satellite in the ITRF coordinate system, determined by the precision orbiting system; $(\Delta X_{ref}\ \ \Delta Y_{ref}\ \ \Delta Z_{ref})^T$ is the fixed offset from the laser emission reference point to the phase center of the GNSS antenna; θ is the laser exit axis pointing angle, i.e., the angle between the projection of the laser emission direction and the XOY plane of this system and the Z-axis negative direction,

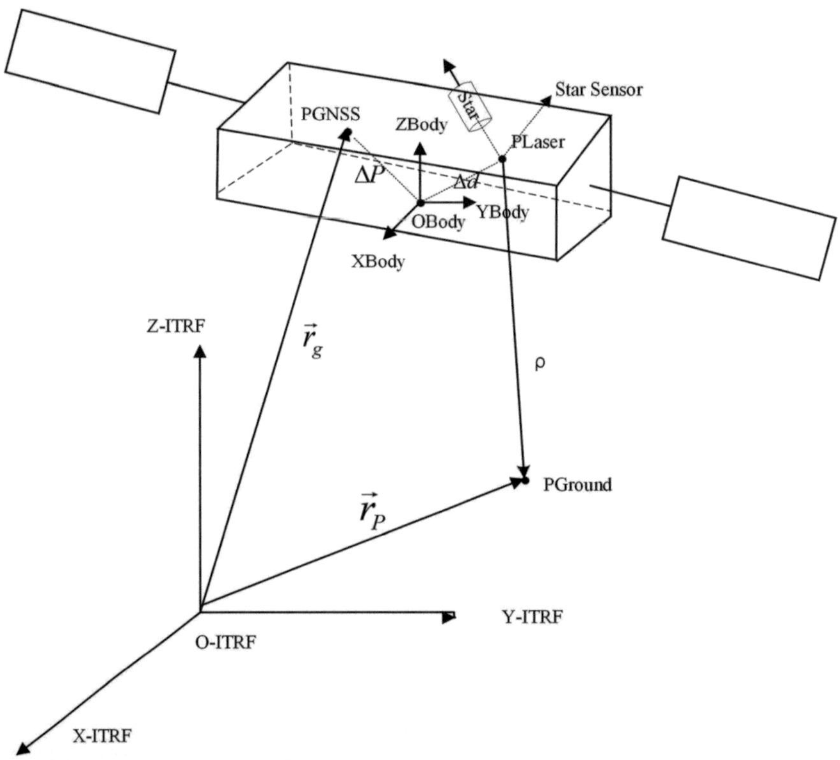

Figure 7.
Schematic diagram of a rigorous geometric positioning model of spaceborne laser point.

α is the angle between the projection of the laser in the XOY plane of this coordinate system and the X-axis positive direction; ρ is the laser range value, corrected by the calibration system error and atmospheric delay correction.

Based on the satellite-based LiDAR measurement of the rigorous geometric model, the positioning and elevation calculation of the laser footprint is achieved by the following process:

1. The measured distance is corrected for systematic errors by in-orbit calibration, and then the corrected distance $\rho' = \rho + \Delta\rho_a$ is considered after the atmospheric delay correction $\Delta\rho_a$. The solution of the atmospheric delay depends on the accuracy of the meteorological data. The measured ground data from the Chinese meteorological station and the meteorological data from the GNSS station are interpolated to the NCEP meteorological station coordinates. Since atmospheric corrections have to be performed in real time, continuous validation work is required during the satellite operation to ensure the accuracy of the atmospheric delay correction values.

2. To effectively eliminate the geolocation error caused by optical line difference, the actual position of the laser footprint on the ground is solved using the satellite attitude at the moment of laser emission and the position of the satellite when the laser reaches the ground. According to the laser launch time tT, the satellite attitude data and the laser pointing measurement data are determined from the precision attitude data, and the pointing correction value obtained from the in-orbit calibration is combined to determine the laser pointing in the ICRF coordinate system. Calculating the three-dimensional coordinates of the laser reference point in the ICRF coordinate system and the rotation matrix of the coordinate system from ICRF to ITRF from the precision orbiting data, based on the laser arrival time tm, combined with the fixed offset of the phase center from the laser emission reference point to the GNSS antenna.

3. Calculate the coordinate vector $\begin{pmatrix} X_{spot} & Y_{spot} & Z_{spot} \end{pmatrix}^T_{ITRF}$ of the laser footprint point in the ITRF coordinate system based on the rigorous geometric model of satellite-based laser altimetry, and then calculate the geodesic coordinates (B, L, H) of the laser footprint point based on the ellipsoidal parameters.

4. Tides include ocean tides, solid tides, etc., whose influence on satellite ranging can reach 0.3–2 m. Therefore, tidal corrections ΔHtide such as ocean tides and solid tides should also be considered to obtain the final coordinates of laser footprints.

Taking ICESat-2 data as an instance, the nominal 6.5 m planimetric positioning accuracy and better than 1 m elevation accuracy were achieved after processing [45]. As shown in **Figure 8**, the evaluation results compared with the airborne DSM data by the iterative least z-difference method in Hanzhong region showed that the horizontal biases of the test data were -0.1 m (east) and -4.1 m (north), and the elevation bias was 0.6 m, which reached the nominal accuracy specifications.

4.2 Photonic point cloud denoising

The extremely high sensitivity of photon-counting LiDAR detection leads to relatively poor data signal-to-noise ratio, and although a narrow-band filter (0.15 nm) is

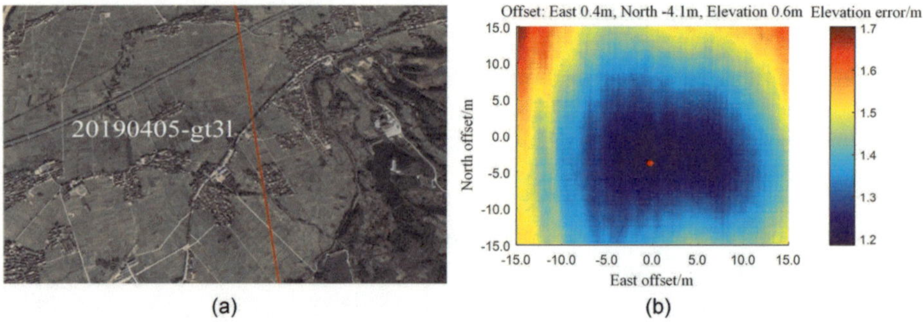

Figure 8.
Accuracy assessment of ICESat-2 using airborne data. (a) Study area; (b) horizontal offset, and red point is best-fitted.

installed, there is still a large amount of background sunlight noise. In some high solar angle and high ground reflectivity scenarios, the background light noise rate reaches about 10 MHz (i.e., 10 million/s, which translates to 1 noise point per 3 m in the elevation direction), so point cloud denoising is critical.

The extremely high sensitivity of photon counting LiDAR detection also leads to a lot of noise and relatively poor signal-to-noise ratio of the data. Although the ICESat-2 receiver is fitted with a narrow-band filter that limits the band range to 532.272 ± 0.15 nm, there is still a large amount of background sunlight in that range. In some high-sun angle and in high ground reflectivity scenarios, the background light noise rate reaches about 10 MHz (i.e., 10 million/s, which translates to 1 noise point per 3 m in the elevation direction) [37], so point cloud denoising is critical.

Most of the currently available photon-counting LiDAR devices record data only along the direction of flight (a swing scan is less common) and therefore are usually processed in a two-dimensional profile. Two denoising algorithms, including the histogram and spatial density methods, are provided in the ICESat-2 basic theory algorithm documents ATL03 [46] and ATL08 [47], respectively: the histogram method considers that the location with the highest number of points in the vertical direction is more likely to be the signal [37]; the spatial density method considers that the signal points are more densely distributed in space, and the density histogram will show the distribution characteristics of "noise on the left, signal on the right" and "high noise and narrow signal" [36]. Based on the "double-peak" distribution of the density histogram, the DRAGANN (Differential, Regressive and Gaussian Adaptive Nearest Neighbor) algorithm is proposed, which uses two Gaussian functions to fit the noise and signal separately and calculates the noise removal threshold adaptively by computing the optimal parameters. The comparison of the two denoising effects is shown in **Figure 9**. The ATL03 algorithm works better in the flat ice cover area, but in the vegetated area, there will be obvious signal point leakage, and ATL08 is more suitable for the vegetated terrain area [48]. In addition, the targeted design on the issues of search kernel shape, terrain correlation, and directional adaptivity can further improve the algorithm performance and obtain better than 98% denoising accuracy [49].

4.3 LiDAR parameters calibration

For a LiDAR measurement satellite similar to ICESat-2 with a 500 km orbital altitude, a pointing angle error of 1 arc second results in a horizontal geolocation error

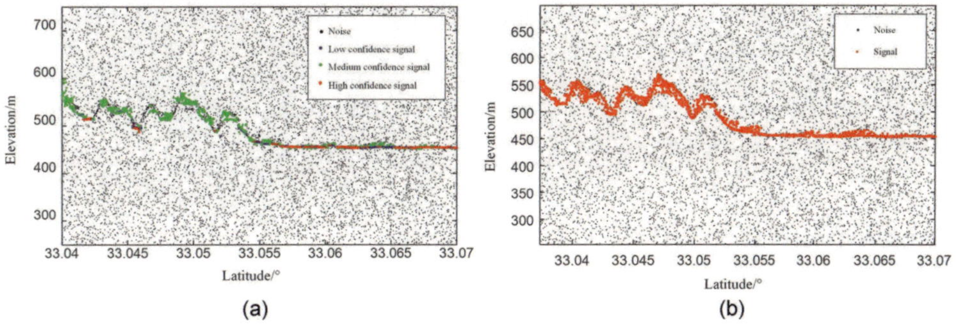

Figure 9.
Comparison of photon counting LiDAR point cloud denoising effects. (a) ATL03 algorithm; (b) ATL08 algorithm.

of about 2.4 m, and an elevation error of 8.3 cm if the ground slope is 2°. Therefore, the rigorous in-orbit calibration of LiDAR work parameters is crucial to the accuracy of elevation measurements.

The on-orbit calibration of satellite-based LiDAR has some similarities with the traditional optical satellite and airborne LiDAR, mainly based on the ground calibration field method or the nature terrain method [50, 51]. The ground-based calibration field directly measures the footprint point through the laser receiver, which has the highest accuracy, but it is necessary to estimate the location of the footprint point and select a suitable location to build the calibration field. The natural terrain method realizes the laser calibration parameter solution by profile alignment with the satellite laser data through the local terrain measured accurately in advance. The rigorous geometric location equation for satellite-based LiDAR is adopted to the basic calibrated model. The iterative pointing angle calibration method was proposed based on the least elevation difference matching criterion. The main steps include two steps: firstly, terrain matching is used to obtain the common terrain feature points whose coordinate deviations are used as input observations, and the second step is to calculate the system calibration parameters by laser beam adjustment. As shown in **Figure 10**, the specific solution can be simplified to calibrate two pointing angular parameters (θ_0, β_0) and 1 range parameter r. It has been shown that the angular calibration accuracy is better than 0.3 arc second using 1 km length laser line matched

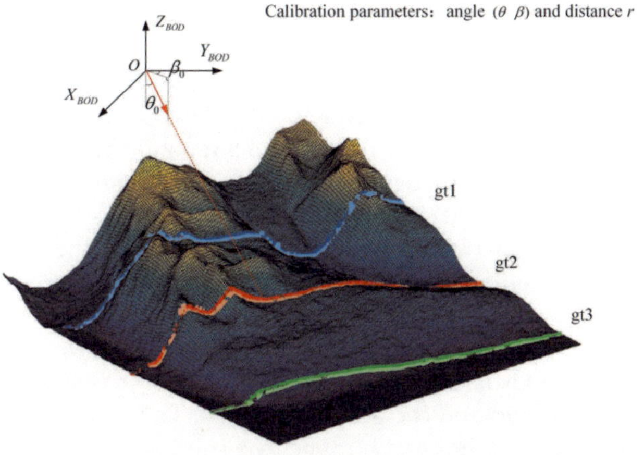

Figure 10.
Diagram of laser on-orbit calibration based on natural terrain.

with high-precision terrain, and the angular calibration accuracy is better than 0.1 arc second when the line length is increased to 2.5 km [52]. In addition, ICESat-2 adopts a similar strategy to ICESat for attitude maneuvering in the oceanic region, and the attitude and distance are calibrated separately by conical scanning, and the long-term drift of the calibrated range values is less than 1 mm/year [53].

5. Application research

5.1 Improve stereo images 3D positioning accuracy

With the characteristics of small ground footprint, high sampling frequency, and large number of beams, the ground observation means of photon counting LiDAR represented by ICESat-2 has greatly improved the planimetric accuracy and flight direction data density of the acquired point cloud, which can be applied as a new 3D control condition to improve the positioning accuracy of satellite images [54–62]. To solve the application problem of photonic point cloud without synchronous image recording plane position, an approach to improve satellite image positioning accuracy with the support of satellite-based photon counting laser point cloud is proposed in this paper; firstly, a 3D terrain profile matching method is used to achieve accurate alignment between photonic point cloud data and DSM automatically generated from satellite stereo image; then terrain feature points are extracted from photonic profile point cloud based on slope change and combined with DSM multiple terrain features to generate common terrain feature control points, and finally introduced into the block adjustment of satellite images with attached parameters as the flat height control condition to further improve the positioning accuracy [63–66]. The experimental results using ZY-3 images and ATLAS ATL03 level data from two regions in Shaanxi Province show that the method can significantly improve the location accuracy without GCPs of satellite images, and compared with the fully uncontrolled positioning and SRTM data-assisted positioning methods, the planimetric and elevation positioning accuracy of ZY-3 image can be further improved by up to 60% and 34%, respectively, which verifies the effectiveness and feasibility of the method [67].

The test data were obtained using ZY-3 satellite images and ICESat-2 ATLAS data in the Xi'an area of Shaanxi Province, China. The ZY-3 satellite image was acquired on April 07, 2019, and the test field contains various types of terrain such as mountains and plains, with elevation relief up to 1200 m. The geographic location of the test area is shown in **Figure 11**, which is a stereoscopic (including front-view, normal-view, and back-view) image area ranging from E108.34°N34.36° to E109.02°N34.92°. The resolution of the front and back view images is 3.5 m, and the normal view image is 2.1 m. The 5-track ATLAS ATL03 data are used, and the acquisition time range is October 26, 2018–January 11, 2021, and the geodetic datum is WGS-84 coordinate system, the plane coordinates are latitude and longitude, and the elevation coordinates are WGS-84 ellipsoidal height, and the distribution on the image is shown in **Figure 11**.

Based on the dense matching using satellite stereo images, the minimum height difference method is used to realize the 3D terrain matching between the photon profiling point cloud and the generated DSM. The basic principle is shown in **Figure 12**, the dashed box is the search range, the solid red-green-blue line is the original three-beam photon point cloud ground track, and the dashed line is the distance search interval of the photon point cloud in its nominated accuracy range.

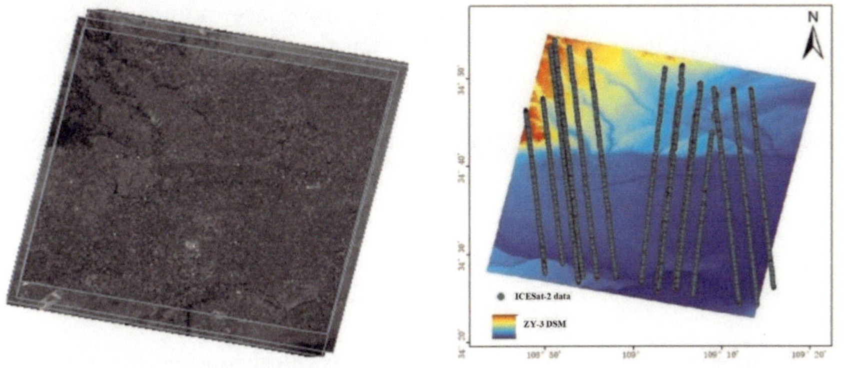

Figure 11.
ZY-3 stereo images and ICESat-2 data.

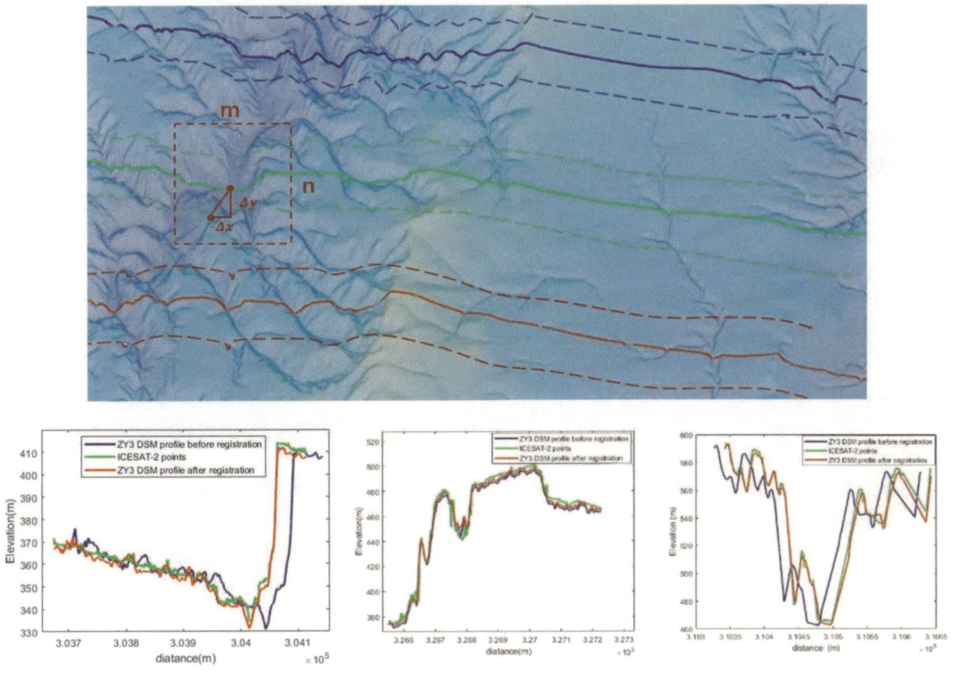

Figure 12.
Diagram of 3D terrain profile matching between multi-beam photon point cloud and DSM generated by stereo images.

The basic method is to first convert the DSM into regular grid data, take one of the tracks of the multibeam laser point cloud strip as the unit, move in the horizontal or vertical direction with a certain search step Δx or Δy, find the same point of the two data plane coordinates, if there is no corresponding point, use its close interpolation, and then calculate the absolute elevation difference between each laser point and the corresponding plane position of the DSM on multiple strips, until traversing all areas in the step setting area to determine the minimum absolute elevation difference position between corresponding points, which is the plane coordinate position of the profile point cloud matched with DSM.

The principle of terrain matching between profiling laser points and DSM based on the minimum elevation difference method is that the absolute elevation difference between the satellite photon point cloud and DSM in the local range is minimum, the accuracy of laser point cloud is high, and the accuracy of DSM is low, and the offset of the two data relative to the plane position is calculated by elevation constraint. The basic equation is shown as follows.

$$d_{\min} = \mathrm{MIN}\sum_{s=1}^{3}\sum_{l=-m}^{m}\sum_{j=-n}^{n}\left(\sum_{i=1}^{N}|Z_i(X,Y)-h_i(X+l\Delta x, Y+j\Delta y)|/N\right) \quad (2)$$

In the formula, s is the number of beams, $d_{\min}$ is the absolute value of the elevation difference between the photon point cloud strip and the DSM data generated from satellite images along the profile, N is the number of laser points in the strip, Z_i is the laser point elevation value, and h_i is the elevation value of the DSM generated from satellite images of the corresponding path. With this condition, the minimum elevation difference alignment can be constructed to obtain the best matching position of the photon profile point cloud strips relative to the direction of the satellite image-generated DSM trajectory, so as to obtain the overall offset of its spatial position.

The slope of the terrain is used as the basic discriminator, which is an important indicator to describe the terrain features and can indirectly reflect the relief pattern and structure of the terrain. For discrete photon point cloud data, a Gaussian fitting method is used to fit the curve, which can obtain more reliable terrain feature points. Then the terrain feature points are extracted from the elevation profile of the satellite-borne photon point cloud tracks based on the slope change, and all the terrain feature points with slope change values larger than the threshold are automatically labeled by setting the slope change threshold as the criterion. Take ICESat-2 point cloud data as an example, the extraction effect is shown in **Figure 13**. The continuous photon point cloud forms the terrain profile, and after the Gaussian curve fitting, it forms a smoother curve.

The principle of laser terrain feature point joint satellite image geolocation is to introduce the laser point as the control condition into the RFM compensation equation, to solve the problem that the parameters to be solved as free unknowns under the uncontrolled condition will lead to unstable accuracy of adjustment [68–75]. The basic process is to firstly match the connection points of satellite images and perform the

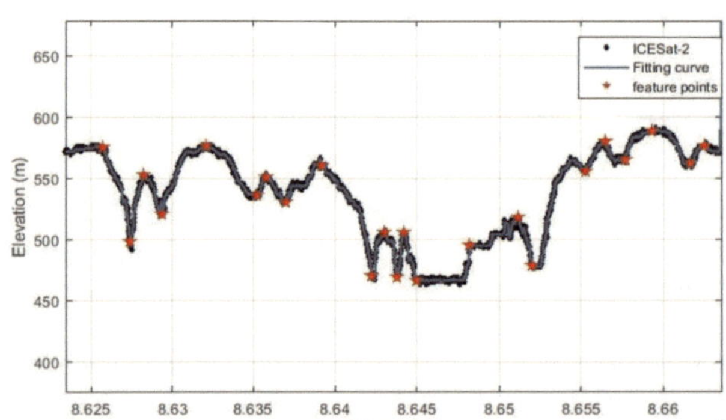

Figure 13.
Terrain feature points extracted from the photon profile point cloud.

free network adjustment; then bring the joint topographic feature points as the plane-height control condition into the block adjustment of satellite images with additional parameters, and different weights be set to participate in the adjustment according to the observation accuracy of different data until the iterative calculation converges; finally, count the adjustment accuracy and output the adjustment results, i.e., the refined image positioning parameters, including RFM model coefficients and additional parameters, which are used together to realize the improvement of satellite image positioning accuracy. The characteristics of this adjustment scheme are that, due to the dense and continuous data of photon point cloud flight, it provides the possibility to automatically extract the topographic features common to both laser and stereo images with active and passive 3D observation data, thus providing an effective way to determine the position of photon point cloud on non-synchronous images, which enables its advantages of both plane and elevation accuracy to be fully exploited.

Comparing the results of uncontrolled, SRTM DEM-assisted and ICESat-2 control points, the uncontrolled block adjustment is based only on the self-contained rational function model, with a plane positioning accuracy of 8.12 m and an elevation accuracy of approximately 8.99 m. By introducing 5 m resolution open-source DOM and open 30 m grid spacing SRTM elevation data, and adding certain accuracy of plane and elevation geometric constraints, the accuracy of SRTM-assisted stereo image positioning is 6.65 m in plane and 2.019 m in elevation. When 1-track ICESat-2 laser point cloud control points are added to the test area, the 3D positioning accuracy of the image can be significantly improved to 2.64 m in plane and 1.39 m in elevation. Compared with the fully uncontrolled positioning and SRTM data-assisted positioning methods, the planimetric accuracy improved from 8.12 m, 6.65 m to 2.56 m, and the elevation accuracy improved from 8.99 m, 2.019 m to 1.319 m, with an improvement of 61.5% and 34.8% respectively compared with SRTM data. In the results from **Table 2**, we can see that the photonic laser point cloud can provide sufficient number of control point data, and the accuracy of auxiliary satellite stereo image positioning can reach the design theoretical accuracy.

The satellite-borne photonic point cloud has new breakthroughs in improving the positioning accuracy of satellite images compared with the existing linear system laser altimetry data and open-source global DEM: first, as a control of the adequacy of data distribution; second, the improvement of data accuracy, especially planimetric accuracy. This all provides a new technology path for achieving global accurate mapping.

Ground control	Image error	L-GCP's number	Check point	Planimetric error (m)	Elevation error (m)
Non-ground control	0.629			8.12	8.99
SRTM	0.330			6.65	2.019
1-Track photon point clouds	0.328	10	5	2.68	1.450
	0.326	18	7	2.64	1.390
5-Track photon point clouds	0.325	10	50	2.57	1.325
	0.325	15	73	2.56	1.319

Table 2.
Accuracy statistics of uncontrolled block adjustment and auxiliary block adjustment.

5.2 Multi-source topographic information fusion

Obtaining high accuracy, high resolution, and global coverage, Digital Elevation Model (DEM) is the goal of aerospace remote sensing mapping [76]. The existing open-source DEM mainly includes SRTM (Shuttle Radar Topography Mission), GDEM (Global Digital Elevation Model), etc. GDEM is an elevation model acquired by Terra satellite optical sensor ASTER with 30 m resolution, covering about 99% of the global land surface. The SRTM is a Space Shuttle radar topographic mapping product with the latest release being the 30 m resolution SRTM-1. GDEM and SRTM-1 have nominal elevation accuracies of 17 m (95% confidence) and 16 m (90% confidence), respectively [77, 78]. In comparison, GDEM has noise and data artifacts due to optical imaging limitations, and SRTM has stable accuracy but data gaps at large slope locations in mountainous areas. ICESat-2 elevation accuracy is much higher than these two open DEMs, while the point density along the track direction is extremely high, and the ICESat-2 point cloud can be used as a control point to correct the open-source DEM elevation deviation and ensure the quality of DEM blank filling [79–82].

Specific processing should pay attention to the unification of coordinate system and elevation datum, ICESat-2 ATL03 point cloud coordinate system is WGS84, the elevation value is ellipsoidal height, the file provides the elevation aberrant value under EGM08 (Earth Gravitational Model 2008) (field/gtx/geophys_corr/geo), which can be read and converted to normal elevation. Corrected DEMs are evaluated with ICESat-2, which allows the mapping of laser points to open-source DEM values to be established and corrections carried out. Vacancy filling is achieved through steps such as rasterization, interpolation (e.g., inverse distance weights, kriging, etc.) [83, 84] and triangulated mesh surface filling. The method that SRTM-1 and GDEM fusion filling were supported by ICESat data has been shown to be effective, and the evaluation shows that the accuracy of GDEM, SRTM-1, and ICESat after fusion processing is 10.9 ± 20.2, 5.8 ± 16.2, and 5.7 ± 14.9 m, respectively, and ICESat-2 is theoretically better for fusion processing.

As shown in **Figure 14**, the results of the fusion processing of multi-source terrain information supported by the strip control datum are used. For the differences of shift and rotation between multiple topographic data in the same area, a fusion processing

Figure 14.
Comparison of the before-after fusion of multi-source topographic data (top view and side view).

method consisting of ICP topographic matching, spatial geometric transformation overall adjustment, and topographic reconstruction is proposed, and the experimental results show that the method can effectively eliminate the systematic geometric errors between different topographic data and improve the fusion accuracy of multi-source topographic data. At the same time, because the terrain matching depends on the terrain features, the feature matching accuracy is better when the terrain is highly rolling, and the overall adjustment accuracy is also better in mountainous or hilly areas compared with flat areas.

5.3 Polar mapping

In recent years, polar ice sheet mass balance studies have revealed that the Antarctic and Greenland ice sheets as a whole are in a state of accelerated melting, which has important implications for both global sea level rise and climate change [85, 86]. Satellite laser altimetry, represented by ICESat-2, is an important tool for mapping changes in polar ice sheet elevation and thus analyzing the ice sheet material balance [87–90].

Mapping polar ice cap elevation changes is one of the main scientific objectives of ICESat-2. At present, the main methods for mapping the elevation change of the polar ice cap surface using ICESat-2 include the intersection point method and the repeated trajectory method. The former uses multiple functions to fit the satellite lift orbits and calculate the ice cap elevation change at the orbit intersection points and their locations. The latter method divides the satellite orbit into kilometer-scale segments and uses least squares to fit the segment function model to calculate the ice cap elevation change. Compared with the traditional single-beam measurement mode used by ICESat, the multi-beam measurement of ICESat-2 greatly increases the number of intersections and effectively improves the ability to obtain details of the ice cap surface elevation change. At the same time, ICESat-2's paired laser beams can independently determine the local surface slope, making it possible to determine surface elevation changes using a single repeated reference track.

ICESat-2 features multiple beams, high repetition frequency, and micro-pulses, which greatly enhance the accuracy and reliability of polar observation results. On the spatial scale of 100 km^2, ICESat-2 has an accuracy of better than 0.25 m/year for ice sheet elevation change measurements and can provide high-precision observations of better than 0.4 cm/year over the entire Antarctic or Greenland ice sheets [53]. **Figure 15** shows the Greenland ice sheet elevation change from 2003 to 2019 based on the intersection of ICESat and ICESat-2. The high-resolution and high-precision measurements provided by ICESat-2 will greatly improve the mapping accuracy in the ice sheet edge areas with large topographic relief and provide more reliable data support for the analysis of the polar ice sheet material balance.

5.4 Three-dimensional topographic mapping

The satellite-based laser survey can quickly and accurately acquire three-dimensional information on the earth's surface, providing a new means for terrestrial elevation three-dimensional mapping. Due to the high altitude of the satellite platform, unlike the airborne platform scanning measurement mode, the satellite-based LiDAR measurement adopts a multi-beam push measurement mode similar to the optical remote sensing satellite line array push imaging. Although the photon counting point cloud data in the flight direction forms a continuous surface elevation profile

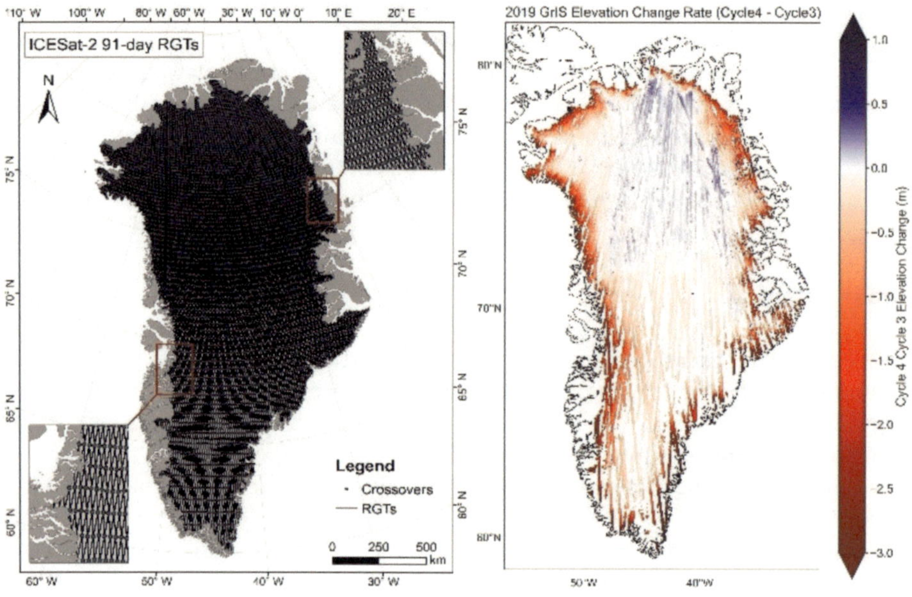

Figure 15.
Changes in the elevation of the Greenland ice sheet from 2003 to 2019 based on the intersection of ICESat and ICESat-2.

due to high sampling frequency, and the ground sampling distance can reach sub-meter level, the sampling interval in the vertical orbit direction may reach hundreds to thousands of meters in magnitude due to the limitation of the number of beams, and the sampling spacing between the two directions is large, making it difficult to effectively measure the object shape. Icesat-2, for example, has a sampling interval of 0.7 meters in the along-track direction and more than 3.3 kilometers in the vertical-track direction, so it is difficult to meet the requirements of three-dimensional terrain mapping with a single pass, which limits the scope of application.

However, as the satellite cycle operation can realize the same area repeatedly observed several times, the sampling distance between sparse beams is gradually reduced, and the point cloud data interval in the vertical track direction can be effectively reduced by using joint processing of multi-track data, **Figure 16** shows the results of multi-track photon counting point cloud data processing using airborne 51-beam LiDAR. The sampling interval between the two directions gradually converges to the same, and finally realizes 3D mapping of the surface object, which can provide a better data source for 3D model reconstruction.

The simulation of satellite operation by airborne experiment shows that although the laser points are widely spaced in the vertical orbit direction, the dense point cloud with more uniform distribution can be obtained by multi-track coverage, which can better reflect the three-dimensional shape of the ground surface. Further, the laser point cloud can be filtered and classified, and a typical algorithm such as the adaptive progressive triangular network filtering method can classify the point cloud into two categories: ground points and non-ground points. Then, according to the spatial distribution of non-ground points, they are further divided into buildings, vegetation and others. In the case of building point clouds, RANSAC plane detection and clustering as well as key point/corner point and boundary extraction are used to achieve automatic reconstruction of

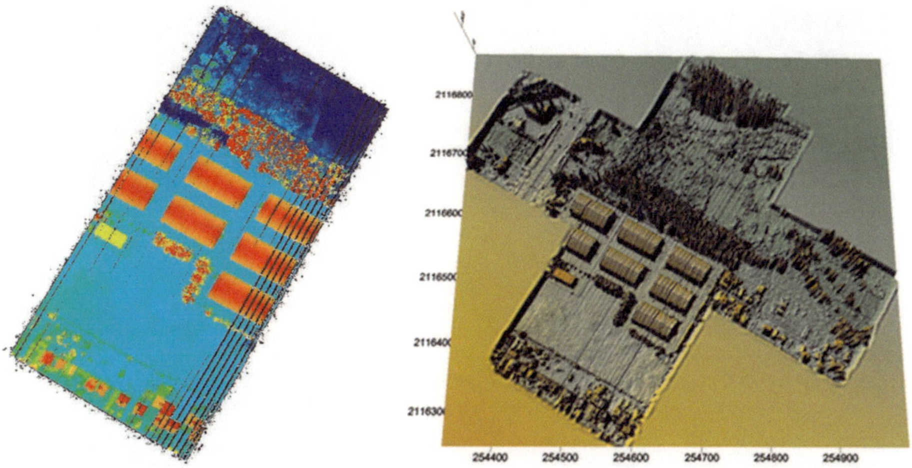

Figure 16.
Airborne multi-orbit photon counting point cloud data and processing results.

Figure 17.
Automatic building extraction from 5m laser point cloud data and LOD1 level 3D reconstruction.

LOD1 level building 3D models, as shown in **Figure 17**, and higher detail levels require further integration of multi-source data such as high-resolution multi-view images, spectral information, geographic information and even mobile phone videos.

At the same time, in order to further improve the efficiency and fineness of the target 3D measurement, the combination of LiDAR measurement and oblique photogrammetry to achieve rapid 3D mapping of the target is an effective technical approach. At present, a payload device Leica CityMapper, which integrates two detection instruments, has emerged in the aerial photogrammetry field, collecting 3D city oblique photography images and laser point cloud data simultaneously, which can effectively improve the productivity and data quality of creating digital city 3D models, realizing efficient and low-cost acquisition of highly detailed and accurate 3D data, and making the widespread use of 3D models a reality. This provides a good reference for the design of a satellite-based hybrid 3D mapping system.

5.5 Shallow Sea topographic mapping

Shallow bathymetry is the basic geographic data required for many applications such as marine engineering construction and marine environmental research, and it is

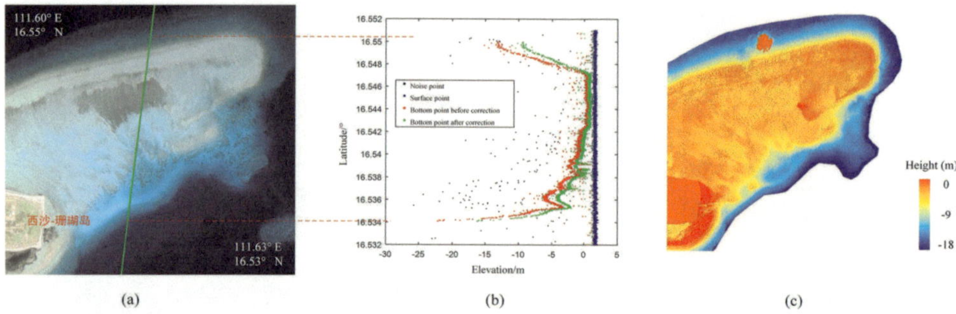

Figure 18.
Schematic diagram of ICESat-2 shallow bathymetry. (a) Geographical location of Xisha Coral Island; (b) ICESat-2 photon profile point cloud and refraction correction effect; (c) shallow water deep pseudo-color map.

a difficult and hot spot in the field of marine mapping to obtain shallow bathymetry accurately and efficiently [91–93].

Laser propagation has to go through a complex process of atmosphere-water body-atmosphere, and the returned laser energy is significantly weakened due to the attenuation of the water body, it is difficult for satellite platforms to detect shallow water depth in the linear detection laser stage, and the ultra-high sensitivity of photon-counting LiDAR makes it a reality. The on-board experimental results of MABEL (ICESat-2 principle prototype) show that the laser bathymetry depth reaches 8 m with a root mean square error 0.7 m [94]; after the release of ICESat-2 data, analysis showed a maximum bathymetric depth of up to 40 m with a root mean square error in the range of 0.43–0.6 m [53]. **Figure 18** shows the joint bathymetry test with WorldView-2 optical images using ICESat-2 at the Xisha coral reef in China, with an accuracy of 0.23 ± 0.98 m. It should be noted that the standard product released by ICESat-2 does not have shallow bathymetry, i.e., the elevation value of ATL03 point cloud does not take into account the refraction of the water body, so the practical application needs to carry out targeted denoising, refraction correction, surface/submerged separation, and bathymetry calculation; at the same time, due to the large spacing of ICESat-2 beam vertical tracks, many distant islands and reefs are not passed by the survey line, so the joint processing of laser and image in a certain range to carry out water body inversion parameter transposition is a more feasible method for the area lacking direct active bathymetry values.

6. Conclusions

The rapid development of satellite-based LiDAR earth observation technology provides an effective way to rapidly collect global surface 3D point clouds, and the advantages and characteristics of laser point cloud data make it a core data component for the construction of the next-generation global mapping database, establishing an infrastructure for the construction of high-precision global control grids and highly efficient surveying of digital elevation maps, and providing support for the rapid construction of high-precision 3D digital geospatial frameworks. At the same time for the global laser point cloud application needs and overall design, it is also necessary to strengthen the research to continuously optimize the technical process, focusing on the following aspects:

1. Global point cloud density. Laser measurements are similar to field measurements in that they are single point measurements, but due to the disorderly nature of laser point cloud data, different applications have corresponding requirements for point cloud density. For regional/national airborne point cloud density is generally 6–10 points/m^2, for example, for the whole of Finland, the area is 338,000 km^2, there are 640 billion points, and the data volume is about 11 TB, while for ground mobile LiDAR point cloud data, the density may exceed 1000 points/m^2. In general, point cloud density is closely related to the application environment. For the global three-dimensional framework construction, the current satellite laser point cloud spacing of 5–10 m should be a reasonable interval, taking into account the mapping efficiency and satellite instrument performance.

2. Point cloud automated mapping. The extraction of multiple spatial information from the satellite-based point cloud data, including terrain, buildings, vegetation, etc., needs to further improve the degree of automation and accuracy, to provide effective support for the construction of three-dimensional digital geospatial information framework. Laser point cloud data automatically classifies ground points and non-ground points, and reliable ground points are the key to rapidly generate digital ground models; non-ground points contain structural information of surface targets, from which there is still a great challenge to automatically reconstruct the 3D framework model of ground targets.

3. Multi-sensor integration. In order to obtain finer surface 3D information, multi-sensor integration and fusion will be an inevitable choice for the development of satellite-based earth observation technology, to achieve the high accuracy of satellite-based LiDAR measurement data and the high resolution of optical/microwave imaging technology combined to give full play to the advantages of multiple detection means [22, 95–97]. The need to study the automatic and accurate matching between LiDAR measurement data and imaging sensor data, as well as the overall fusion strategy for the difference in accuracy between the two, are issues that need to be further studied and solved.

4. Rapid application mode. Point cloud data are a discrete expression of the earth surface, directly superimposed orthophoto fast 3D display has a more intuitive visualization ability than only using vector maps, on the basis of which further superimposed thematic information can meet the rapid application needs in emergency situation. This mode has higher efficiency and flexibility compared with point cloud 3D model reconstruction. Therefore, it is necessary to pay attention to the development of technologies such as 3D game engines, including the host platform, development environment, and tools applicable to global point cloud data, as well as the construction and development of standard interfaces.

Since the first satellite-based laser altimetry satellite ICESat-1 was successfully launched in 2003, after nearly two decades of development, satellite LiDAR earth observation technology has achieved great progress and gradually developed into an emerging remote sensing exploration instrument. Especially since 2018, ICESat-2 satellite (2018), GEDI (2018), Gaofen-7 (2019), Gaofen-14 (2020), and terrestrial ecological environment monitoring carbon satellite (2022) have been continuously launched into the sky, and the satellite laser observation technology has come to a climax of development. The update of acquiring tools has provided new opportunities for the global mapping technology, and studies have shown that spaceborne laser observation data have strong application potential in the fields of precise positioning of global earth, polar ice sheet monitoring, and shallow ocean topography mapping, etc. It is foreseen that in the near future, with the further development of satellite-based LiDAR technology, it will become possible to directly acquire global digital surface models and digital elevation models, thus providing an active technical way to establish uniformly accurate elevation data in a global even grid, and a new means to establish a higher precision 3D digital earth model.

Author details

Fang Yong*, Zhang Li, Gong Hui, Cao Bincai*, Gao Li and Hu Haiyan
Xi'an Research Institute of Surveying and Mapping, Xi'an, China

*Address all correspondence to: yong.fang@vip.sina.com and cbcontheway@163.com

References

[1] Li DR. From geomatics to geospatial intelligent service science. Acta Geodaetica et Cartographica Sinica. 2017;**46**(10):1207-1212

[2] Juho-Pekka V, Antero K, Harri K, etc. Nationwide point cloud-the future topographic Core data. ISPRS International Journal of Geo-Information. 2017;**6**(8):243

[3] Li R, Wang C, Su GZ, et al. Development and application of spaceborne lidar. Science & Technology Review. 2007;**14**:58-63

[4] Yu ZZ, Hou X, Zhou CY. Development status of spaceborne laser altimetry technology. Laser & Optoelectronics Progress. 2013;**50**(02): 52-61

[5] Fang Y, Cao BC, Gao L, et al. Development and application of lidar mapping satellite. Infrared and laser engineering. 2020;**49**(11):19-27

[6] Zhu XX, Wang C, Xi XH, et al. Research progress on data processing and application of icesat-2 spaceborne photon counting lidar. Infrared and Laser Engineering. 2020;**49**(11):76-85

[7] Fang Y. Research and Practice of Spaceborne SAR Image Spatial Information Extraction Technology. Zhengzhou: Zhengzhou University of Information Engineering; 2001

[8] Fang Y, Gong H, Zhang L, Haiyan H. From global laser point cloud acquisition to 3D digital geospatial framework: The advanced road of global accurate mapping. Laser & Optoelectronics Progress. 2022;**59** (12):1200002 (in Chinese)

[9] Xin X, Liu B, Di K, Yue Z, Gou S. Geometric quality assessment of Chang'e-2 global DEM product. Remote Sensing. 2020;**12**(3):526-546

[10] He Y. Research on Lunar Topography Mapping Technology Based on Lunar CCD Image and Laser Altimetry Data. Zhengzhou: PLA University of Information Engineering; 2012

[11] Hu H, Wu B. Planetary 3D: A photogrammetric tool for 3D topographic mapping of planetary bodies. The ISPRS Annals of the Photogrammetry, Remote Sensing and Spatial Information Sciences. 2019; **IV-2/W5**:519-526. DOI: 10.5194/isprs-annals-IV-2-W5-519-2019

[12] Guoyuan L. Earth Observing Satellite Laser Altimeter Data Processing Method and Engineer Practice. Wu Han: Wu Han University; 2017 (in Chinese)

[13] Sun G, Ranson KJ, Kharuk VI, et al. Validation of surface height from shuttle radar topography mission using shuttle laser altimeter. Remote Sensing of Environment. 2003;**88**(4):401-411

[14] Simard M, Pinto N, Fisher J, et al. Mapping forest canopy height globally with spaceborne lidar. Journal of Geophysical Research. 2011, 2011; **116**(G4):1-12

[15] Scott BL, Tim R, Taylor T, et al. Algorithm Theoretical Basis Document (ATBD) for GEDI Waveform Geolocation for L1 and L2 products. [EB/OL]. 2019. Available from: http://lpdaac.usgs.gov/documents/579/GEDI_WFGEO_ATBD_v1.0.pdf [Accessed: October 24, 2022]

[16] Xinming T, Junfeng X, Xinke F, et al. ZY3-02 laser altimeter on-orbit

geometrical calibration and test. Acta Geodaetica et Cartographica Sinica. 2017;**46**(6):714-723 (in Chinese)

[17] Chen GD, Wang P, Zhao W. ICESat ice sheet height measurement accuracy estimation based on intersection mismatch. Journal of Surveying and Mapping Science and Technology. 2018; **35**(03):226-230

[18] Hu GJ, Fang Y, Zhang L. Development of spaceborne lidar and analysis of surveying and mapping application prospect. Surveying and Mapping Technology and Equipment. 2015;**17**(02):34-37

[19] Chen BX, Zhu Q, Hu H. ZY-3 global DEM accuracy evaluation method based on glas laser altimetry data. Geographic Information World. 2019;**26**(06):70-73

[20] Wei ZQ. Pay attention to ICESat. In: Proceedings on the Progress of Geodesy and Geodynamics. HuBei Science and Technology Press; 2004. pp. 41-44

[21] Li GY, Tang XM, Zhang CY, et al. Multi-criteria constraint algorithm for selecting ICESat/GLAS data as elevation control points. Journal of Remote Sensing. 2017;**21**(1):96-104

[22] Wang J, Zhang Y, Zhang ZX, et al. ICESat laser points assisted block adjustment for mapping Satellite-1 stereo imagery. Acta Geodaetica et Cartographica Sinica. 2018;**47**(3): 359-369

[23] Zhang ZX, Tao PJ. An overview on "cloud control" photogrammetry in big data era. Acta Geodaetica et Cartographica Sinica. 2017;**46**(10): 1238-1248

[24] Tang XM, Xie JF, Mo F, et al. GF-7 dual-beam laser altimeter on-orbit geometric calibration and test verification. Acta Geodaetica et Cartographica Sinica. 2021;**50**(3): 384-395

[25] Aiyan G, Jun D, Chengguang Z, et al. Design and on-orbit validation of GF-7 satellite laser altimeter. Spacecraft Engineering. 2020;**39**(3):43-48 (in Chinese)

[26] Leigh HW, Magruder LA, Carabajal CC, et al. Development of onboard digital elevation and relief databases for ICESat-2. IEEE Transactions on Geoscience & Remote Sensing. 2015; **53**(4):2011-2020

[27] Gong P, Huang HB. The application prospect of lidar technology in topographic mapping in China. Geographic Information World. 2008; **6**(06):45-48

[28] Cao BC, Fang Y, Gao L, et al. Verification of ICESat-2/ATLAS laser altimetry data accuracy by airborne point cloud detection. Journal of Surveying and Mapping Science and Technology. 2020;**37**(01):50-55

[29] Abdalati W, Zwally HJ, Bindschadler R, et al. The ICESat-2 laser altimetry mission. Proceedings of the IEEE. 2010; **98**(5):735-751

[30] Ma Y. Data Processing and Error Analysis of Spaceborne Laser Altimetry System. Wuhan: Wuhan University; 2013

[31] Li X, Liao H, Zhao ML, et al. Research on LiDAR surveying satellite detection capacity for different terrains. Acta Geodaetica et Cartographica Sinica. 2014;**43**(12):1238-1244

[32] Li S, Xiao JM, Ma Y, et al. Study on atmospheric refraction delay correction model of spaceborne laser altimetry

system. Optics and Optoelectronics Technology. 2013;**11**(01):7-11

[33] Yi H, Li S, Ma Y, et al. On orbit calibration of laser altimeter based on footprint detection. Acta Physica Sinica. 2017;**66**(13):118-126

[34] Li M. Research on Photon Counting Lidar Technology Based on fiber Optics. Beijing: University of Chinese Academy of Sciences; 2017

[35] Guo Y. Research on Key Technologies of Photon Counting 3D imaging Lidar. Beijing: Graduate School of Chinese Academy of Sciences; 2011

[36] Cao BC, Fang Y, Gao L, et al. Implementation and accuracy evaluation of ICESat-2 ATL08 denoising algorithms. Bulletin of Surveying and Mapping. 2020;(5):25-30 (in Chinese)

[37] Neumann TA, Martino AJ, et al. The ice, cloud, and land elevation Satellite-2 mission: A global geolocated photon product derived from the advanced topographic laser altimeter system. Remote Sensing of Environment. 2019; **233**:111325

[38] Dong J, Ni W, Zhang Z, Sun G. Performance of ICESat-2 ATL08 product on the estimation of forest height by referencing to small footprint LiDAR data. National Remote Sensing Bulletin; **25**(6):1294-1307

[39] Sun W, Jin J, Li G, Yao J. Accuracy evaluation of laser altimetry satellite ICESat-2 in monitoring water level of Taihu lake. Science of Surveying and Mapping. 2021;**46**(11):6-11

[40] Zhu X, Wang C, Xi X, Nie S, Li D. Research progress of ICESat-2/ATLAS data processing and applications. Infrared and Laser Engineering. 2020; **49**(11):76-85

[41] Neuenschwander AL, Magruder LA. Canopy and terrain height retrievals with ICESat-2: A first look. Remote Sensing. 2019;**11**(14):1721

[42] Bentley Pointools: Point-Cloud Processing Software. Available from: https://www.bentley.com/en/products/product-line/reality-modeling-software/bentley-pointools [Accessed: October 24, 2022]

[43] Euclideon Udstream. Available from: https://www.euclideon/udstream/ [Accessed: October 24, 2022]

[44] Li G, Ye F, Tang X, et al. Influence of range gate width on detection probability and ranging accuracy of single photon laser altimetry satellite. Journal of Geodesy and Geoinformation Science. 2020;**3**(2):36-44

[45] Cao B, Fang Y, Gao L, Hu H, Jiang Z. Verification of ICESat-2/ATLAS laser altimetry data accuracy using airborne point cloud. Journal of Geomatics Science and Technology. 2020;**37**(01): 50-55

[46] Neumann T, Brenner A, Hancock D, et al. NASA. ICE, CLOUD, and Land Elevation Satellite (ICESat-2) Algorithm Theoretical Basis Document (ATBD) for Global Geolocated Photons ATL03 [EB/OL] 2019. Available from: https://icesat-2.gsfc.nasa.gov/sites/default/files/page_files/ICESat2_ATL03_ATBD_r001.pdf [Accessed: 25 August 2020]

[47] Neuenschwander A, Pitts K. ICE, CLOUD, and Land Elevation Satellite (ICESat-2) Algorithm Theoretical Basis Document (ATBD) for Land - Vegetation Along-track products (ATL08) [EB/OL]. 2019. Available from: https://icesat-2.gsfc nasa.gov/sites/default/files/page_files/ICESat2_ATL08_ATBD_r001_0.pdf [Accessed: 25 August 2020]

[48] Neuenschwander A, Pitts K. The ATL08 land and vegetation product for the ICESat-2 Mission. Remote Sensing of Environment. 2019;**221**:247-259

[49] Feng X, Gui Y, Rong S, et al. An adaptive directional filter for photon counting lidar point cloud data. Journal of Infrared, Millimeter, and Terahertz Waves. 2017;**36**(1):107-113 (in Chinese)

[50] Xie J, Liu R, Mei Y, et al. Preliminary pointing bias calibration of ZY3-03 laser altimeter. Journal of Geodesy and Geoinformation Science. 2021;**4**(3): 91-100

[51] Tang X, Li G, Gao X, Chen J. The rigorous geometric model of satellite laser altimeter and preliminarily accuracy validation. Acta Geodaetica et Cartographica Sinica. 2016;**45**(10): 1182-1191

[52] Nan YM, Feng ZH, Liu EH, et al. Iterative pointing angle calibration method for the spaceborne photon-counting laser altimeter based on small-range terrain matching. Remote Sensing. 2019;**11**(18):2158

[53] Markus T, Neumann T, Martino A, et al. The ice, cloud, and land elevation Satellite-2 (ICESat-2): Science requirements, concept, and implementation. Remote Sensing of Environment. 2017;**190**:260-273

[54] Marco GMS. Automated geometric correction of high-resolution Pushbroom satellite data. Photogrammetric Engineering & Remote Sensing. 2008; **74**(1):107-116

[55] Reinartz P et al. Orthorectification of VHR optical satellite data exploiting the geometric accuracy of TerraSAR-X data. ISPRS Journal of Photogrammetry and Remote Sensing. 2011;**66**(1):124-132

[56] Zhou P, Tang X, Cao N, Wang X, Li G, Zhang H. SRTM-aided stereo image block adjustment without ground control points. Acta Geodaetica et Cartographica Sinica. 2016;**45**(11):1318-1327

[57] Di K, Liu B, Peng M, et al. An initiative for construction of new-generation lunar global control network using multi-mission data. ISPRS—International Archives of the Photogrammetry, Remote Sensing and Spatial Information Sciences. 2017; **XLII-3/W1**:29-34

[58] Cao H, Tao P, Li H, Shi J. Bundle adjustment of satellite images based on an equivalent geometric sensor model with digital elevation model. ISPRS Journal of Photogrammetry and Remote Sensing. 2019;**156**:169-183

[59] Wang M, Wei Y, Yang B, Zhou X. Extraction and analysis of icesAT-2/ ATLAS global elevation control points. Geomatics and Information Science of Wuhan University. 2021;**46** (2):184-192

[60] Hanley HB, Yamakawa T, Fraser CS. Sensor orientation for high-resolution satellite imagery. In: International Archives of Photogrammetry and Remote Sensing. No. 34. 2002

[61] Toutin T. Spatiotriangulation with multisensor VIR/SAR images. IEEE Transactions on Geoscience and Remote Sensing. 2004;**42**(10):2096-2103

[62] Liu S, Lv Y, Tong X, et al. An alternative approach for registration of high-resolution satellite optical imagery and ICESat laser altimetry data. Sensors. 2016;**16**(12):2008

[63] Li G, Tang X. Analysis and validation of ZY-3 02 satellite laser altimetry data.

Acta Geodaetica et Cartographica Sinica. 2017;**46**(12):1939-1949

[64] Li G. Earth observing satellite laser altimeter data processing method and engineer practice. Acta Geodaetica et Cartographica Sinica. 2018;**47**(12): 1691-1691

[65] Li G, Tang X, Gao X, et al. Improve the ZY-3 height accuracy using icesat\glas laser altimeter data. The International Archives of the Photogrammetry, Remote Sensing and Spatial Information Sciences. 2016;**XLI-B1**:37-42

[66] Li G, Wang Y, et al. ZY-3 block adjustment supported by glas laser altimetry data. Photogrammetric Record. 2016;**31**(153):88-107

[67] Lai WY, Fang Y. Block adjustment of satellite images supported by spaceborne photon point cloud. Laser & Optoelectronics Progress. 2023;**60**(10): 1028013

[68] Grodecki J, Dial G. Block adjustment of high-resolution satellite images described by rational polynomials. Photogrammetric Engineering & Remote Sensing. 2003;**69**(1):59-68

[69] Yang B, Wang M, Pi Y. Block-adjustment without GCPs for large-scale regions only based on the virtual control points. Acta Geodaetica et Cartographica Sinica. 2017;**46**(7):874-881

[70] Li G, Tang X, Wang H, et al. Research on the ZY-3 block adjustment supported by the GLAS laser altimetry data. In: The 3rd Annual Academic Conference on High Resolution Earth Observation. 2014. pp. 586-600

[71] Wang J, Zhang Y, Zhang Z, et al. ICESat laserpoints assisted block adjustment for mapping satellitle-1 stereo imagery. Acta Geodaetica et Cartographica Sinica. 2018;**47**(03): 359-369

[72] Cao N, Zhou P, Wang X, et al. Refined processing of laser altimeter data-aided satellite geometry model. Journal of Remote Sensing. 2018;**22**(4):599-610

[73] Tang X, Liu C, Zhang H, Wang X, Guoyuan L, Mo F, et al. GF-7 satellites stereo images block adjustment assisted with laser altimetry data. Geomatics and Information Science of Wuhan University. 2021;**46**(10):1423-1430

[74] Cai C. Research on ICESat-2 Point Cloud Data Assisted Satellite Image DEM Extraction Method. XiAn: Chang'an University; 2020

[75] Zhang L, Xing S, Xu Q, Zhang G, Li P, Jiao L, et al. Joint block adjustment for ATLAS data and ZY3-02 stereo imagery. Infrared and Laser Engineering. 2020; **49**(S2):155-162

[76] Han H, Ding YL, Zhu Q, et al. Precision global DEM generation based on adaptive surface filter and poisson terrain editing. Acta Geodaetica et Cartographica Sinica. 2019;**48**(3): 374-383

[77] Li ZH, Li P, Ding D, et al. Research progress of global high resolution digital elevation models. Geomatics and Information Science of Wuhan University. 2018;**43**(12):1927-1942

[78] Yue L, Shen H, Zhang L, et al. High-quality seamless DEM generation blending SRTM-1, ASTER GDEM v2 and ICESat/GLAS observations. ISPRS Journal of Photogrammetry and Remote Sensing. 2017;**123**:20-34

[79] Hossein A, Peter R. Accuracy enhancement of ASTER global digital elevation models using ICESat data. Remote Sensing. 2011;**3**(7):1323-1343

[80] Yang S, Yang N, Chen C, et al. Accuracy assessment and improvement of SRTM DEM based on ICESat/GLAS under the consideration of data coregistration over Jiangxi Province. Journal of Geo-information Science. 2021;**23**(5):869-881

[81] Chen B, Zhu Q, Hu H. Global DEM accuracy evaluation method based on GLAS laser altimetry data. Geomatics World. 2019;**26**(6):70-73

[82] Zhang T, Cen M, Feng Y, Yang R, Ren Z. DEM matching algorithm using least trimmed squares estimator. Acta Geodaetica et Cartographica Sinica. 2009;**38**(2):144-151

[83] Shan J, Deng F, Tao PJ, et al. Crowdsource Image Photogrammetry [M]. Beijing: Science Press; 2019. pp. 49-73

[84] Wang X, Holland DM, Gudmundsson GH. Accurate coastal DEM generation by merging ASTER GDEM and ICESat/GLAS data over Mertz glacier, Antarctica. Remote Sensing of Environment. 2018;**206**: 218-230

[85] Li Q, Ke C, Zhang J, Xiaoyi S. Estimation of the mass balance of Greenland IceSheet from 2003 to 2019 based on ICESat and ICESat-2 laser altimetry data. Journal of Geo-information Science. 2022;**24**(3): 558-571

[86] Shepherd A, Ivins E, Rignot E, et al. Mass balance of the Antarctic ice sheet from 1992 to 2017. Nature. 2018; **558**(7709):219-222

[87] Gao XM, Li GY, Guo JQ, et al. Discussion on the development of three pole observation laser altimetry satellite. Infrared and Laser Engineering. 2020; **49**(11):40-46

[88] Luthcke SB, Rowlands DD, Williams TA, et al. Reduction of ICESat systematic geolocation errors and the impact on ice sheet elevation change detection. Geophysical Research Letters. 2005; **32**(21):1-4

[89] Brunt KM, Neumann TA, Smith BE. Assessment of ICESat-2 ice sheet Surface Heights, based on comparisons over the interior of the Antarctic ice sheet. Geophysical Research Letters. 2019; **46**(22):13072-13078

[90] Csatho BM, Schenk AF, Van Der Veen CJ, et al. Laser altimetry reveals complex pattern of Greenland Ice Sheet dynamics. Proceedings of the National Academy of Science of the United States of America. 2014;**111**(52):18478-18483

[91] Cao B, Fang Y, Gao L, Haiyan H, Jiang Z, Sun B, et al. An active-passive fusion strategy and accuracy evaluation for shallow water bathymetry based on ICESat-2 ATLAS laser point cloud and satellite remote sensing imagery. International Journal of Remote Sensing. 2021;**42**(8):2783-2806. DOI: 10.1080/ 01431161.2020.1862441

[92] Cao BC, Fang Y, Jiang ZZ, Gao L, Hu HY. Shallow water bathymetry from WorldView-2 stereo imagery using two-media photogrammetry. European Journal of Remote Sensing. 2019;**52**(1): 506-521

[93] Parrish CE, Magruder LA, Neuenschwander AL, et al. Validation of ICESat-2 ATLAS bathymetry and analysis of ATLAS's bathymetric mapping performance. Remote Sensing. 2019;**11**(14):1634

[94] Forfinski-Sarkozi N, Christopher P. Analysis of MABEL bathymetry in Keweenaw Bay and implications for ICESat-2 ATLAS. Remote Sensing. 2016; **8**(9):772

[95] Du QY, Mai XM, Chu FX, et al. Registration of unconventional aerial images and airborne lidar data based on ICP algorithm. Remote Sensing Information. 2014;**29**(06):16-20

[96] Cao BC, Fang Y, Jiang ZZ, et al. Bathymetry of icesat-2 laser satellite and optical remote sensing image fusion. Marine Surveying and Mapping. 2020; **40**(05):21-25

[97] He H, Wang ZM, Wen HF. Research on tilt correction of spaceborne SAR interferometric DEM using satellite laser altimetry data. Remote Sensing Information. 2009;**04**:85-88

Chapter 10

Modeling the Use of LiDAR through Adverse Weather

Maria Ballesta-Garcia, Gerard DeMas-Giménez and Santiago Royo

Abstract

Due to the outstanding characteristics of LiDAR imaging systems, they seem essential for the consolidation of novel applications related to computer vision, in fields such as autonomous vehicles, outdoor recognition, and surveillance. However, the final technology implementation still has some uncertainties and needs in-depth work for its use in these real-world applications. Under the presence of adverse weather conditions, for example in fog, LiDAR performance is heavily influenced and the quality of the detection becomes severely degraded. The range is reduced due to the dispersion of the media and the sensor could be saturated due to backscattering or deliver a very limited range. Light propagation modeling through turbid media is used as a tool to understand and study these phenomena. Mie Theory allows the characterization of the optical media and light-particle interactions. Monte-Carlo methods are used to solve the radiative transfer problem related to these situations. When working with those models, the results obtained are in accordance with the ones shown in experimental tests, and it is possible to predict the necessities and problems of the designed systems.

Keywords: LiDAR, turbid media, scattering, Mie theory, radiative transfer problem, Monte-Carlo method, light propagation modeling, pulsed light

1. Introduction

One of the up-to-date new applications of LiDAR technology is its use in transport, surveillance, and security [1]. A pulsed laser is used to measure ranges (variable distances). Thanks to distance calculations, it is possible to perform 3D mapping using different approaches [2] and to recover the geometry of the scene and not just a projection as in a conventional camera.

LiDAR technology has been presented as a disruptive technology regarding computer vision, as it gives precise and real-time visualization of the surrounding area and its distribution of objects. It offers an outstanding performance toward the required specifications [3]. However, the feasibility of the instrument for day-to-day outdoor uses is still facing numerous questions, one of them related to its detection breakdown when working in adverse weather conditions [4–6].

The propagation of light through scattering media such as fog, rain, smoke, dust, or others shows two main problems: the attenuation of the pulse and the saturation of the sensor [7–9]. The first one is a result of the dispersive effect and absorption characteristics of the media, which leads to the loss of energy while pulse light is propagating. The second is related to the backscattering that light undergoes when just entering the media, which may blind the sensor [10]. Due to their nature, both phenomena can be easily studied with models if the optical properties of the media are known. When working with models, the results obtained are in accordance with the ones shown in experimental tests.

At this point, it is worth showing in **Figure 1** a point cloud obtained under fog conditions. The corresponding RGB image of the scene without fog is shown along three different views of the point cloud (with fog). Objects found in the scene are indicated on the RGB image with red letters. All point clouds are labeled as in the RGB image to facilitate its interpretation, i.e., the same letters are used inside black tags on the point clouds to point to objects. Z is the direction of propagation, normal to the RGB image and along the tunnel forming the fog chamber, Y refers to the height of the chamber and X to the width, distances are shown in meters [m]. On the left, we

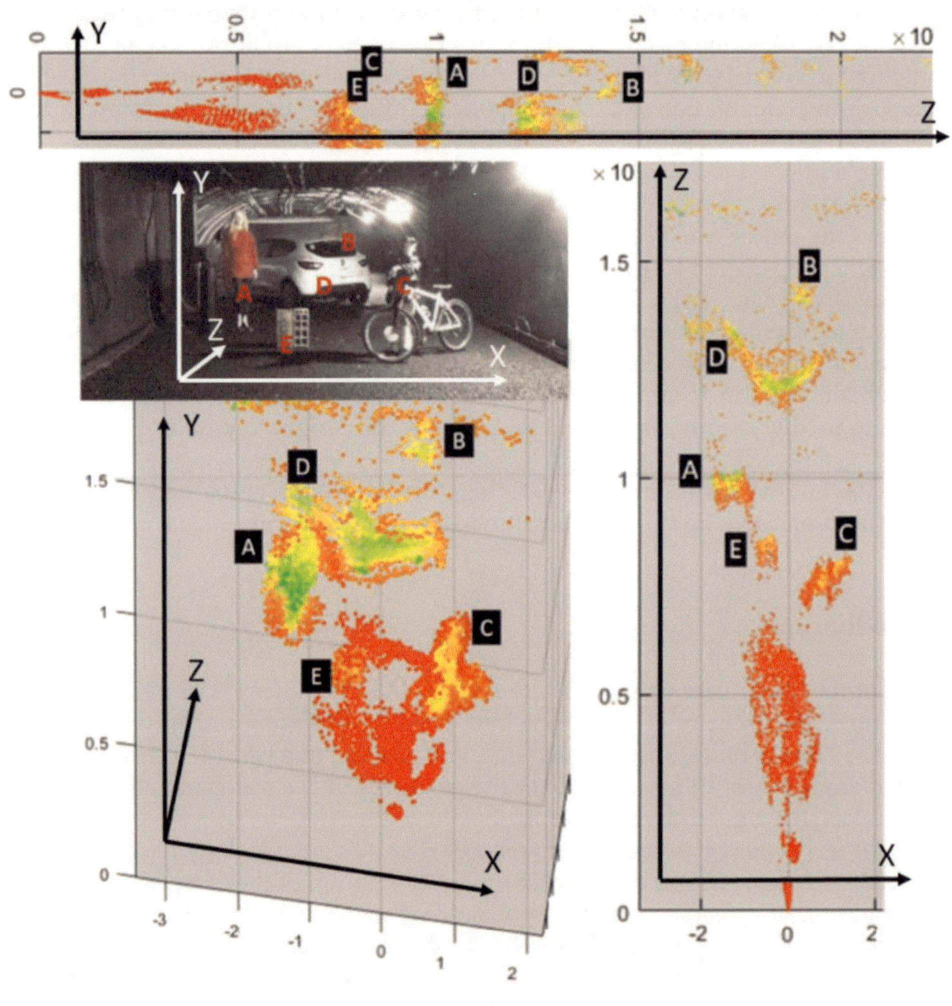

Figure 1.
Different views of a point cloud under the presence of artificial fog. The scene used is shown in the RGB image without fog.

present a 3D view selected with an adequate orientation to highlight the aspect of the point cloud. On top, there is a YZ view, i.e., a side view of the scene; on the right there is a ZX view, which is a top view of the scene. These views are useful to notice the spread of the point cloud around a determined object/distance, and the points appearing due to the backscattering of fog, especially just in front of the sensor. In conclusion, the point cloud is rather noisy and the range is limited.

Nowadays, there is no current solution to overcome the problem presented. The described effects are still present and seriously damage the performance of LiDAR systems in adverse weather conditions. However, there are several lines of research trying to find ways to improve this situation [11–16]. The challenge of this topic makes researchers work with the novelties in optical engineering (optical design, materials of the components, new sensors ...) along with the basis of physics of light (propagation of light, light-matter interactions ...) [1].

In this chapter, we want to review some basics of the physics of light to properly become aware of the problem. How does light propagate through any turbid media? How is the media characterized? Which is its effect when working with pulsed light? If one wants to face this problem with a plausible solution, it is necessary to know and understand the involved physical phenomena in depth. Modeling allows us to go deeper into what is happening. Thus, it will be also reviewed how models are conventionally approached and which is the state of the art on the topic.

2. Optical properties

2.1 Description of the optical properties

For the description of the physical phenomenon of propagation of a pulse of light through a turbid media (energy transfer), the balance of incoming, outgoing, absorbed, and emitted photons is used. This is a wide, established, and well-known field, with a dense literature corpus [17–23].

Adverse weather conditions can be thought of as turbid media, as particles of different types and shapes (water, smoke, dust ...) are found suspended within the main media, which is air [24]. By definition, a turbid medium is characterized by having localized non-uniformities randomly distributed within it. These optical non-uniformities are usually inclusions of one substance within another with a different index of refraction n. These inclusions cause the medium to be optically inhomogeneous and cause it to behave as a scattering media.

In the air, which is a non-absorbing media, the inclusions, widely represented as "particles," are in charge of the actual absorption of part of the propagating energy and the actual change of direction of light. As a result, the dominant effects in the medium are absorption and scattering. According to these two effects, the medium is represented by several key optical parameters: the absorption coefficient μ_a, the scattering coefficient μ_s, the scattering phase function $p(\theta, \phi)$, and the asymmetry factor g; which respectively describe: the absorbing and scattering power, the probability of scattering in a particular direction (θ, ϕ) of the media, and the degree of scattering in the forward direction.

An absorbing medium is composed of particles (or other structures) that can absorb light and transform it into its internal energy as the beam is propagated along the medium, which results in a gradual reduction of the light intensity. To characterize this phenomenon, one uses a parameter that is called the absorption coefficient μ_a (in units of $length^{-1}$, usually in cm^{-1}), which quantifies the absorbing effect of the medium.

The deviation of light from its straight trajectory due to localized non-uniformities in the medium is known as scattering. The particles become scattering centers, which, when exposed to light, modify the electromagnetic field and re-emit it in a different direction. Analogously to the absorption case, a parameter called the scattering coefficient μ_s (units of $length^{-1}$, typically cm^{-1}) is defined, which quantifies the scattering effect.

When a collimated beam of light passes through a volume of the medium, it will lose intensity due to both processes: absorption and scattering. In general, both processes cannot be distinguished. This effect is characterized by what is known as the extinction or attenuation coefficient $\mu_t = \mu_a + \mu_s$ (cm^{-1}). The extinction coefficient measures the total loss of a narrow-beam intensity, i.e. the loss due to absorption and the loss corresponding to the part of photons that have not been scattered in the forward direction.

Along with the scattering coefficient, a scattering event needs other parameters to be completely defined. In particular, photons may not be isotropically scattered and may need to have this dependence characterized. To completely define the deflection of the trajectory in space after a scattering event, two angles are used. The deflection scattering angle θ (which ranges from 0 to π) defines the deflection of the trajectory in the scattering plane—the plane formed by the direction of the incident light and the direction of the outgoing scattered light, i.e., the cone angle; and the azimuthal angle ϕ(which is defined from 0 to 2π) defines the change in the plane perpendicular to the scattering plane. These angles are shown in **Figure 2**, where the geometry of a scattering event is schematically depicted.

The directionality of the scattering effect is quantified using a phase function $p(\theta, \phi)$. The phase function corresponds to the angular distribution of the light scattered by a scattering center at a given wavelength. It can be thought of as a

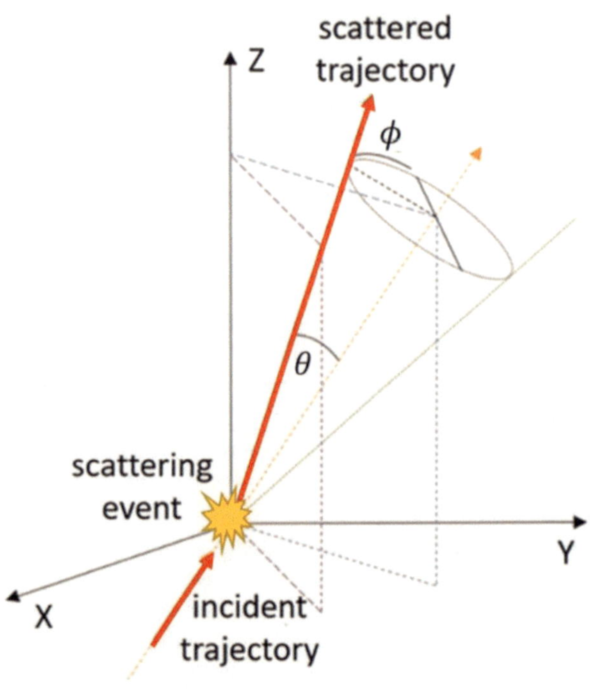

Figure 2.
Schematics of the scattering geometry.

probability density function, showing the chances of a photon being scattered in a particular direction.

When the suspended particles in media have no preferred scattering orientation (spherical symmetry), it is known as an isotropic medium. Then, light is scattered equally in all directions. However, it is more frequent that natural materials scatter light preferentially in the backward or forward direction. For those non-isotropic cases, it is interesting to know the amount of energy retained in the forward/backward direction after a single scattering event. If a photon is scattered so that its trajectory is deflected by a certain deflection angle θ, then the component of the new trajectory, which is aligned in the former forward direction, is presented as $\cos(\theta)$. The mean value of this cosine is known as the anisotropy factor g, and it is defined as:

$$g \equiv \langle cos\,\theta \rangle = \int_0^{\pi} p(\Theta)\, cos\,\Theta \cdot 2\pi sin\Theta\, d\Theta \tag{1}$$

Its value varies in the range from -1 (total backward scattering) to 1 (total forward scattering), being $g = 0$ the value corresponding to isotropic scattering.

In addition, scattering may be elastic and inelastic. Without entering into many details, elastic scattering is associated with an interaction with no energy losses (and, thus, no wavelength change), while inelastic scattering corresponds to a process with energy transfer and thus to a wavelength shift. Generally speaking, elastic scattering is the predominant effect when propagating through turbid media, as approximately only one in each 10^7 photons is inelastically scattered. Elastic interactions between photons and scattering particles are mainly described using two physical models: the Rayleigh and the Mie theories. The description of the process using one or another model is linked to the particle size and the wavelength of the incident light. However, Mie Theory is a general model developed using Maxwell equations and gives exact solutions in all cases, which means that it could be valid for any particle size.

Generally speaking, a turbid medium is described as a system of discrete spherical particles suspended within a base medium, which is exact for the case of fog (water suspended in air) and an approximation for the rest of the cases mentioned for the atmosphere (smoke ...). Such spherical particles enable Mie Theory to be used for its characterization. This theory gives quantitative results of the interaction of an electromagnetic plane wave with a single homogeneous sphere, being likely the most important exactly soluble problem in the theory of absorption and scattering by small particles. Mie Theory allows the calculation of the absorption and scattering coefficients and the phase function of a spherical particle of radius a as a function of the incident radiation wavelength λ, the size parameter $x = 2\pi a/\lambda$, and the complex refractive indexes of the particles and the host material. The derivation of the complete theory may be long and tedious, and detailed information can be found in [19, 25].

2.2 Mie theory

The interaction of light with a spherical particle can be described and quantified using Mie Theory. Some conditions, however, have to be fulfilled to apply the theory. It has to be supposed that the media is homogeneous and that the particles that are embedded within it are spherical, homogeneous, and act independently—so they are distant enough from each other to consider only far-field scattering effects. Their radius and refractive index need to be known. As with most problems in theoretical optics, the scattering of light by a homogeneous sphere is treated as a formal problem of Maxwell's theory with the appropriate boundary conditions [19, 25].

Suppose that one or more particles are placed in a beam of electromagnetic (EM) radiation. The rate at which EM energy is received by a detector downstream from the particles is denoted by U. If the particles are removed, the power received by the detector is U_0, where $U_0 > U$. We say that the presence of the particles has resulted in the extinction of the incident beam. If the medium in which the particles are embedded is non-absorbing (such as air), the difference $U_0 - U$ accounts for absorption and scattering by the embedded particles (water droplets). Although the specific details of extinction depend on many parameters, certain general features are shared in common by all particles.

Now, consider extinction by a single arbitrary particle embedded in a non-absorbing medium and illuminated by a plane wave. If an imaginary sphere of radius r is constructed around the particle, the net rate at which EM energy crosses the surface A of the sphere is W_A. If $W_A < 0$, energy is absorbed within the sphere (being W_A the rate at which energy is absorbed by the particle).

W_A maybe conveniently written as the sum of:

$$W_A = -W_S + W_{ext} \tag{2}$$

W_S is the rate at which energy is scattered across the surface A, and, therefore, W_{ext} is just the sum of the energy absorption rate and the energy scattering rate: $W_{ext} = W_S + W_A$.

Now it is possible to define C_x as the ratio of W_x (being x: A, S or ext) to I_i (incident irradiance):

$$C_x = \frac{W_x}{I_i} \tag{3}$$

The C_x quantities are called cross sections of the particle, and they have area dimensions. Let the total energy scattered in all directions be equal to the energy of the incident wave falling on the area C_S; likewise, the energy absorbed inside the particle may be defined as the energy incident in the area C_A, and the energy removed from the original beam may be equal to the energy incident in the area C_{ext}, which gives an idea of the amount of energy removed from the incident field due to scattering and/or absorption generated by the particle. The law of conservation of energy then requires that:

$$C_{ext} = C_A + C_S \tag{4}$$

When solving Maxwell's equations for the defined problem, the scattered EM field is written as an infinite series in the vector spherical harmonics M_n and N_n, which are the EM normal modes of the spherical particle. Thus, the scattered field is expressed as a superposition of these normal modes, each weighted by the appropriate coefficient a_n or b_n, known as scattering coefficients.

It is found that:

$$C_S = \frac{W_s}{I_i} = \frac{2\pi}{k^2}\sum_{n=1}^{\infty}(2n+1)\left(|a_n|^2 + |b_n|^2\right) \tag{5}$$

$$C_{ext} = \frac{W_{ext}}{I_i} = \frac{2\pi}{k^2}\sum_{n=1}^{\infty}(2n+1)\, Re\,\{a_n + b_n\} \tag{6}$$

$$C_A = C_{ext} - C_S \tag{7}$$

Assuming that the series expansion of the scattered field is uniformly convergent, it is proved that the series can be terminated after:

$$n = x + 4x^{1/3} + 2 \tag{8}$$

where $x = ka$ is the size parameter, with k being the wavenumber and a the radius of the sphere.

We also need to obtain explicit expressions for the scattering coefficients:

$$a_n = \frac{S'_n(y)S_n(x) - mS_n(y)S'_n(x)}{S'_n(y)\zeta_n(x) - mS_n(y)\zeta'_n(x)} \tag{9}$$

$$b_n = \frac{mS'_n(y)S_n(x) - S_n(y)S'_n(x)}{mS'_n(y)\zeta_n(x) - S_n(y)\zeta'_n(x)} \tag{10}$$

where:

$$S_n(z) = \sqrt{\frac{\pi z}{2}} J_{n+0.5}(z) \tag{11}$$

$$\zeta_n(z) = \sqrt{\frac{\pi z}{2}} H^{(2)}_{n+0.5}(z) \tag{12}$$

With $J_{n+0.5}(z)$ being the half-integral-order spherical Bessel function of first kind and $H^{(2)}_{n+0.5}(z)$ the half-integral-order Hänkel function of the second kind. The variables x and y, in this case, correspond to $x = ka$ and $y = mka$, and m is the relative refractive index between the sphere and the medium in which it is embedded; and finally, $S'_n(z)$ and $\zeta'_n(z)$ denote the derivatives of the corresponding functions.

Once the cross section of a single interaction has been computed, one needs to characterize the media. μ_a can be understood as the sum of contributions of the absorption cross sections of the absorbers per unit volume, i.e., the product of the absorption cross section $C_A(cm^2)$ by the density of absorbers ρ_a $(\#/cm^{-3})$:

$$\mu_a = C_A\rho_a \tag{13}$$

Analogously, knowing the number of scattering particles per unit volume ρ_s $(\#/cm^{-3})$ and their C_S, it is possible to compute the scattering coefficient μ_s of the propagating medium as:

$$\mu_s = C_S\rho_s \tag{14}$$

The phase function and the asymmetry factor can also be computed using Mie Theory:

$$p(\theta) = \frac{2\pi}{k^2 C_S}\left(|S_1|^2 + |S_2|^2\right) \tag{15}$$

$$g = \frac{2\pi}{k^2 C_S}\left(\sum_{n=1}^{\infty}\frac{2n+1}{n(n+1)} Re\left\{a_n b_n^*\right\} + \sum_{n=1}^{\infty}\frac{n(n+2)}{n+1} Re\left\{a_n a_{n+1}^* + b_n b_{n+1}^*\right\}\right) \tag{16}$$

being S_1 and S_2:

$$S_1 = \sum_{n=1}^{\infty} \frac{2n+1}{n(n+1)} (a_n \pi_n \cos\theta + b_n \tau_n \cos\theta) \tag{17}$$

$$S_2 = \sum_{n=1}^{\infty} \frac{2n+1}{n(n+1)} (b_n \pi_n \cos\theta + a_n \tau_n \cos\theta) \tag{18}$$

From which a_n and b_n are the coefficients computed in Eqs. (9) and (10), and π_n and τ_n two angle-dependent functions known as Mie angular functions. These angular functions are generated with the associated Legendre polynomials, and they can be calculated from the recurrence relations:

$$\pi_n = \frac{(2n-1)\mu}{n-1} \pi_{n-1} - \frac{n}{n-1} \pi_{n-2} \tag{19}$$

$$\tau_n = n\mu\pi_n - (n+1)\pi_{n-1} \tag{20}$$

where $\mu = \cos\theta$, and the first terms of π_n are $\pi_0 = 0$ and $\pi_1 = 1$.

3. Propagation model

Propagation through turbid media is characterized by multiple scattering. This is the situation in which along the total traveling distance L of light, each photon undertakes on average many collisions with particles of the medium. Under these conditions, $\mu_s L \geq 1$ is fulfilled and light properties such as phase, polarization, and ray trace are severely degraded, so only radiometric information may be considered. The basic theory that allows the calculation of light distributions in multiple scattering media with absorption is the radiation transfer theory (RTT). Its core is the radiation transfer equation (RTE)—linear transport or Boltzmann equation, a balance equation characterizing the flow of photons (or any particle) in a given volume element [17].

Some examples of multiple scattering media are biological tissue, nebulous media, colloids, murky water, clouds, and also adverse weather such as fog and rain. For adverse weather conditions, particles suspended in the air act as light scatterers, being responsible for the reduction of the visibility near the ground surface. By using Mie Theory and knowing the approximate distribution of the size particles [26] (for example, modified gamma distributions are the standard choice for fitting the models of fog droplet size distributions [27]), it is possible to give values to the optical parameters (μ_a, μ_s, $p(\theta, \phi)$, and g) and characterize the medium for solving the RTE.

Once the media has been characterized, the RTE has to be solved to describe the energetic distribution of light. However, the analytical solution to this equation is complicated and often impossible to solve when boundaries, inhomogeneities, or nonstationary effects are involved. The Monte-Carlo (MC) method is used as the conventional tool to arrive at a statistical solution in these cases. MC is a stochastic method, which offers robustness and versatility for solving this kind of problem. By modeling, we can predict the shape, range, and intensity of a pulsed LiDAR working through turbid media.

The MC method refers to a technique first proposed by Metropholis and Ulam [28] to simulate physical processes using a stochastic model. Regarding the radiative transfer problem, the MC method is based on recording photons' histories as they are scattered and absorbed, using the global optical properties of the medium [29, 30].

Simulations show the expected movement of individual photons, which are treated as particles of light that move according to certain probability density functions. The photon moves in a straight path and may come across obstacles. At the surface of the obstacles, it may undergo absorption or scattering. Then, the photon continues its flight until it is absorbed or leaves the medium (some examples are shown in Refs. [31–33]). When this process is repeated for many individual photons, MC simulations provide a flexible approach towards light transport that yields maps of the light distributions in turbid media induced by a light source.

MC solutions can be obtained for any desired accuracy, which is proportional to $1/\sqrt{N}$ where N is the number of photons propagated. Thus, relative errors less than a few tenths of a percent will require the propagation of substantial numbers of photons (between 10^6 and 10^9 photons) and may require large amounts of computer time [33].

3.1 Model structure

The modeling of pulsed light propagation is based on solving the RTE using MC method [29–35]. The scheme shown in **Figure 3** summarizes the whole code structure:

Next, we will briefly describe the main steps of the MC method.

3.1.1 Photon initialization

There are different MC approaches. Variance reduction techniques are used to reduce the number of photons necessary to achieve the desired accuracy for a Monte-Carlo calculation. One of them is implicit capture. In this approach, to improve the efficiency of the MC program, many photons (a packet) are propagated along each pathway.

The MC method thus begins by launching a packet of photons, with a size called weight, into the medium. Usually, one may think that only one photon follows each pathway, and at each step, the photon may be either absorbed or scattered. The packet approach is used to improve the efficiency of the MC program, as many photons (a packet) are propagated along each pathway at the same time. If a packet of photons followed each pathway, then some portion of the packet would be absorbed in each step. So, after each propagation step, w is reduced by the probability of absorption.

When a collimated beam normally incident on a slab vertically is simulated, the packets' (which from now on will be called photons) initial direction is chosen downward into the medium, orthogonal to the surface. The initial coordinates of the photon are usually identical for all photons and the weight of the photon is initially set to unity.

3.1.2 Propagation distance

Once launched, the photon moves a distance Δs. The most efficient method is to choose a different step-size Δs for each photon step. The probability density function for the step size follows Beer's law, so the probability of scattering is proportional to $e^{-\mu_t \Delta s}$, being $\mu_t = \mu_a + \mu_s$ - and is chosen in such a way that it is the distance at which the photon interacts with the obstacle. The Δs value may be generated with this probability density function as a function of a random number ξ, uniformly distributed between 0 and 1:

$$\Delta s = \frac{-\ln \xi}{\mu_a + \mu_s} \tag{21}$$

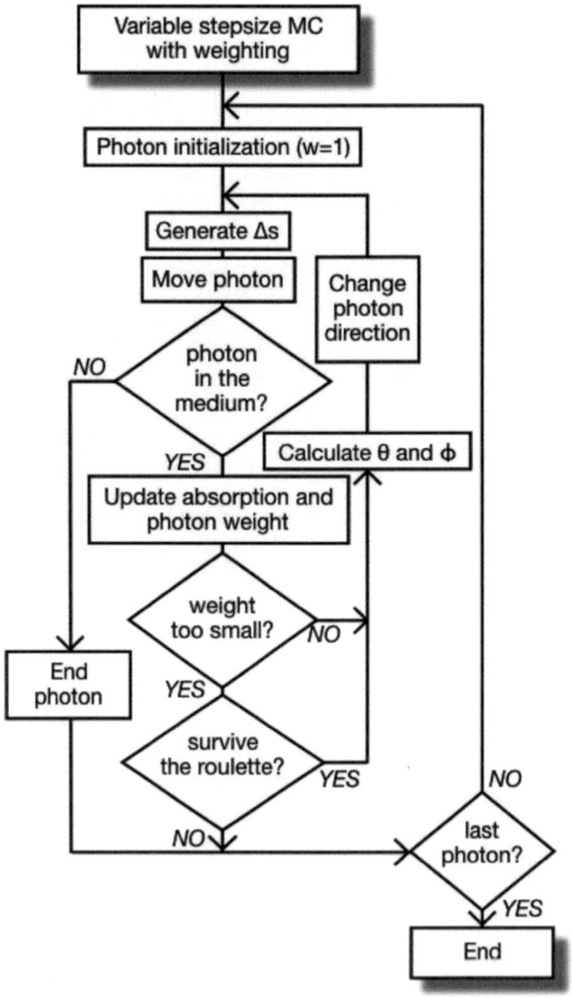

Figure 3.
Flux diagram of the MC method used.

3.1.3 Moving the photon

A photon is uniquely described by five variables: three spatial coordinates for the position and two directional angles for the direction of travel. However, it is convenient to describe the photon's spatial position with three Cartesian coordinates $(x;y;z)$ and the direction of travel with three direction cosines (u_x,u_y,u_z), corresponding to the cosine of the angle that the photon's direction makes with each axis: X-, Y-, and Z-axis respectively. In this case, the required formulas for propagation are simpler. For a photon located at $(x;y;z)$ traveling a distance Δs in the direction (u_x,u_y,u_z), the new coordinates $(x';y';z')$ are given by:

$$\begin{aligned} x' &= x + u_x\Delta s \\ y' &= y + u_y\Delta s \\ z' &= z + u_z\Delta s \end{aligned} \tag{22}$$

3.1.4 Absorption

There exist different methods to consider absorption during propagation. The most followed approach is known as the technique of implicit capture. It assigns a weight to each photon as it enters the medium. After each propagation step, the photon is split into two parts: a fraction is absorbed, and the rest is scattered. Thus, in every encounter, the photon interacts with the scatters. Upon interaction, a fraction μ_a/μ_t—that corresponds to the probability of absorption—if the photon's weight is absorbed and the remaining μ_s/μ_t fraction of the photon's weight is scattered and continues to propagate. The absorbed fraction is placed in a bin of the MC absorption matrix ($\boldsymbol{W}$), which encloses the current photon position:

$$\boldsymbol{W}(x, \mathrm{y}, \mathrm{z}) = \frac{\mu_a}{\mu_t} w \tag{23}$$

and the new photon's weight (w') is updated:

$$w' = \frac{\mu_s}{\mu_t} w \tag{24}$$

As the weight of the photon falls below a certain threshold, a photon termination strategy (see 3.1.6) needs to be implemented.

3.1.5 Scattering

The angle at which the photon is bent when it strikes an obstacle is defined by the two angles of scattering (θ and ϕ) and the normalized phase function, which describes the probability density function for the angles at which a photon is scattered. As has been studied in previous sections, each type of obstacle is characterized by a different phase function.

For isotropic scatterers, such as spherical particle approximation, the phase function has no azimuthal dependence and only depends on θ. Thus, ϕ is uniformly distributed between 0 and 2π and may be generated by multiplying a random number uniformly distributed over the interval 0 to 1 by 2π. Using Mie Theory, it is possible to compute the phase function as shown in Eq. (15), which then has to be sampled to compute θ.

Nevertheless, an approximation with a lower computational cost is close enough for most cases. In practice, it is more convenient to use semiempirical approximations of the scattering phase function, with much lower computational cost. This choice is a compromise between realism and mathematical tractability. The most common example is the Henyey-Greenstein (HG) phase function, which includes the average cosine g:

$$p^{HG}(\theta) = \frac{1}{4\pi} \frac{1 - g^2}{\left(1 + g^2 - 2g\ cos\,\theta\right)^{3/2}} \tag{25}$$

HG is an analytical function originally derived for modeling scattering by interstellar dust and is widely used in biomedical optics and other fields. Then, one only has to invert the probability density function to obtain the generating function of θ:

$$cos\,\theta = \frac{1}{2g}\left[1+g^2 - \left(\frac{1-g^2}{1+g+2g\xi}\right)^2\right] \tag{26}$$

where ξ is a random number uniformly distributed between 0 and 1.

If a photon is scattered at an angle (θ,ϕ) from the initial direction (u_x,u_y,u_z) in which it is traveling, then the new direction $(u_x{}',u_y{}',u_z{}')$ is specified by:

$$u'_x = \frac{sin\,\theta}{\sqrt{1-u_z^2}}\left(u_x u_z\,cos\,\phi - u_y\,sin\,\phi\right) + u_x\,cos\,\theta \tag{27}$$

$$u'_y = \frac{sin\,\theta}{\sqrt{1-u_z^2}}\left(u_y u_z\,cos\,\phi + u_x\,sin\,\phi\right) + u_y\,cos\,\theta \tag{28}$$

$$u'_z = -\,sin\,\theta\cos\phi\sqrt{1-u_z^2} + u_z\,cos\,\theta \tag{29}$$

3.1.6 Photon termination

A photon will be terminated either if it exits the considered space or if it is absorbed. However, using the technique of implicit capture, photon weight will never be a mathematical 0. Thus, a photon is terminated when its weight falls below a given threshold, despite the termination of the process using a threshold introduces a systematic negative bias into the system regarding energy conservation. To reduce this bias, the Roulette method is used [36].

The Roulette method works in the following way. A predefined number N between 2 and 10 is usually chosen in practice. Once the weight of the photon reduces below a sufficiently small threshold, a uniform random number ξ between 0 and 1 is generated. The photon is removed from the system only if $\xi > 1/N$. The photon that survives is continued with its current weight increased by a factor of N. The result is that photons are usually terminated, but energy is conserved by the occasional surviving photon being given extra weight. Since millions of photons are run, the statistically averaged result is correct.

3.1.7 Observable

Once all the photons have been run, the data are stored in the bin of $\boldsymbol{W}$, as $W(x,y,z)$, in units of weight/bin. In order to obtain physical values, some changes are needed. The results will be delivered in the form of fluence rate Φ $[Watts/cm^2]$.

Firstly, we need to convert from weight to photons. Then we normalize it by the appropriate voxel volume (V) and the total number of photons, which leads to the photon absorption density in units of cm^{-3}. Finally, the fluence rate Φ [Watts/cm^2] is obtained by dividing the power density absorbed by the absorption coefficient and multiplying it by the incident power P $[Watts]$:

$$\Phi(x,y,z) = \frac{W(x,y,z)\cdot N_p \cdot P}{V\cdot N_t\cdot \mu_a} \tag{30}$$

In which N_p is the number of photons per packet and N_t the total number of photons.

3.1.8 Temporal approach

Given that pulsed LiDAR is time-dependent, we are interested in time-resolved simulations. MC propagating code is easily modified by adding a record of the time history of each photon in order to have the time-tracking of light propagation. By using this history, it is possible to generate the temporal profile of optical power. The time that each package remains in space is obtained by dividing the length of the path traveled by the speed of light in that medium [30, 37].

In summary, modeling the radiative transfer problem in scattering media involves a potentially simple methodology: a photon packet is emitted, it travels a distance, and something happens to it that affects its energetic weight in successive events. MC method gives us a statistical solution, which becomes a powerful tool to design and characterize our systems.

4. Applications

Models provide a tool to predict light behavior in different situations. There exist two basic analytical models that correspond to two particular situations in which the RTE has an analytical solution. The solutions can be used to compare the results obtained with stochastic models, such as MC, and verify that MC can be also applied to solve RTT without solving the RTE.

On the one hand, it is well known that when absorption prevails over scattering ($\mu_a \gg \mu_s$), Beer-Lambert law can be applied [22]. On the other hand, under certain conditions of the diffusion regime ($10\mu_a \approx \mu_s$), in source-free and homogeneous media conditions, there is an analytical solution based on the theory of photon density wave in steady conditions [38]. **Figure 4** shows, in both cases, results obtained from MC simulations (red dots) and the analytical solution (blue line). They are superimposed showing the validity of our implementation of the MC approach.

As discussed, if one wants to study pulsed light propagation of a LiDAR system, time-resolved simulations have to be used. The MC code needs to be modified by adding a record of the time history of each photon to have the time-tracking of light propagation. By using this history, it is possible to generate the temporal profile of

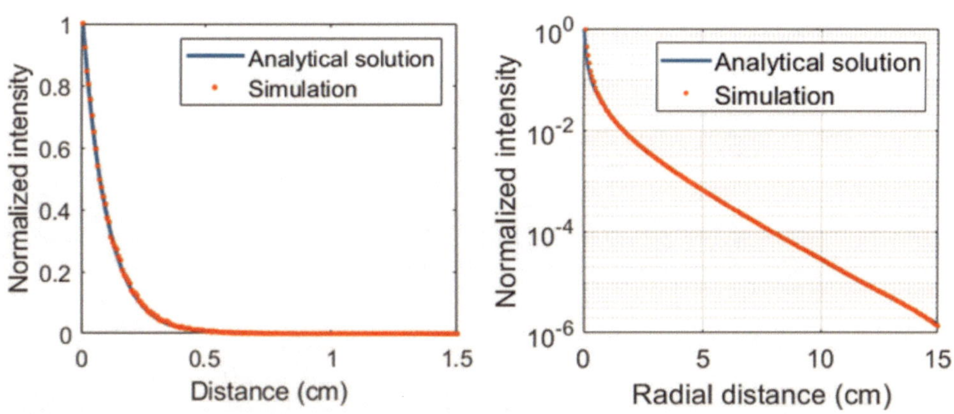

Figure 4.
Left: Beer-Lambert Law. Propagation of light through a medium with $\mu_a = 10cm^{-1}$,$\mu_s = 0.05cm^{-1}$,$g = 0.9$.
Right: Diffusion regime. Propagation of light through a medium with $\mu_a = 0.05cm^{-1}$,$\mu_s = 20\ cm^{-1}$,$g = 0.2$.

optical power. For this case, there also exist situations in which there is an analytical solution to the RTE [39]. For example, in the case of an infinite diffusing medium with a point source, as shown in **Figure 5**, the setting between the analytical and the simulated solution is almost perfect.

Finally, **Figure 6** shows how models based on MC methods are able to predict pulsed light interactions, which is, in fact, the topic of interest in this chapter. It shows the temporal profile of a pulse of light sent of Gaussian shape and 1 mJ of energy (in orange) and light received back to the lighting (in blue). The medium is characterized by the following properties: $\mu_a = 0\ cm^{-1}$,$\mu_s = 0.01\ cm^{-1}$,$g = 0.9$, which would correspond to an almost clear media, and a 100% reflective plane is placed at a distance of

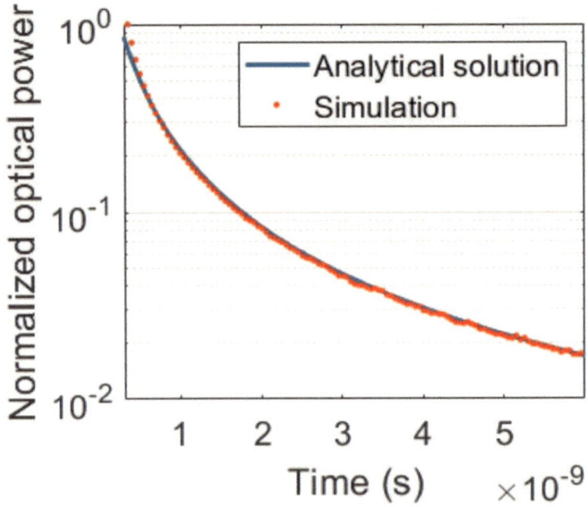

Figure 5.
Time-resolved propagation of light through a medium with $\mu_a = 0cm^{-1}$,$\mu_s = 2cm^{-1}$,$g = 0.9$.

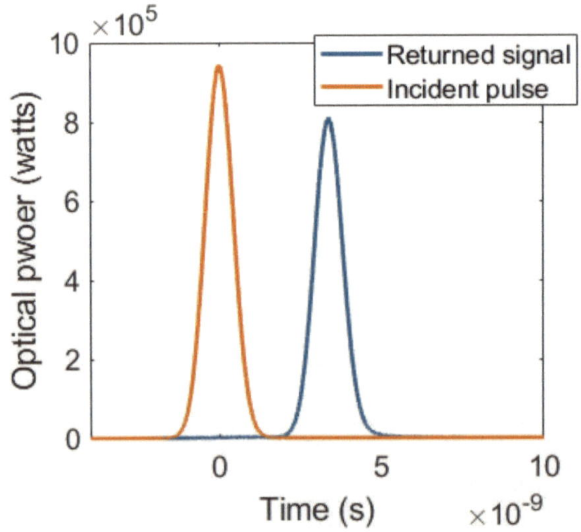

Figure 6.
Light signal sent and received through a medium with $\mu_s = 0.01cm^{-1}$,$g = 0.9$.

0.5 m. It can be observed from this simple figure the working principle of LiDAR. The position of the reflected light peak, with respect to the initial pulse, allows us to obtain information on the position of the object. The shape of the pulse is maintained and the integral of both pulse profiles meets the expected energy of the propagation calculated from the range equation.

One of the problems shown by LiDAR technology when working through adverse weather conditions is the saturation of the sensor. This saturation is the result of the backscattering that light undergoes when just entering the media, and it may saturate

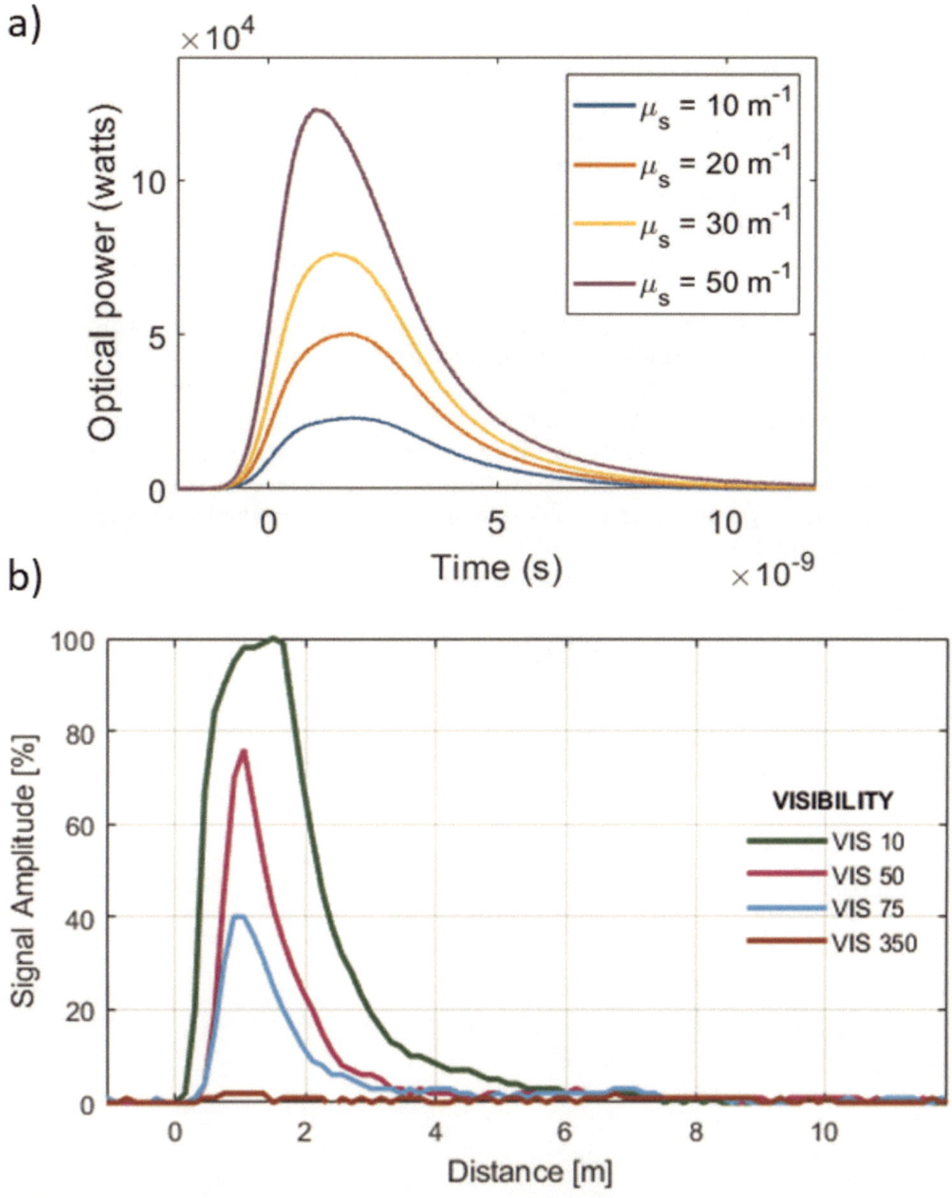

Figure 7.
a) Detected backscattering in a medium with g = 0.9 and different μ_s, for a light pulse with an initial energy of 1 mJ. b) Experimental results of detected backscattering in a foggy medium for different visibilities.

the sensor. Thus, to start with, it is interesting to analyze simple backscattering interactions in different media in order to have an idea of how the first light that would arrive at our detection in foggy conditions would influence our sensor. As the final objective is to use these MC methods to improve a long-range active sensing technique that interacts with a target, we also focused on studying the effect of reflected light that reaches back the plane of illumination, i.e., the light that comes back from an object hidden behind the turbid environment. Therefore, we want to know how the pulse may be attenuated and what the range of the system will be.

Figure 7a shows different profiles of backscattered light simulated using an MC code. The greater the scattering coefficient, the greater the amount of backscattered energy and the more extended in time. It can be also observed that the backscattering signal does not have the same shape as the initial pulse, which is defined as a perfect Gaussian-shaped pulse [40]. **Figure 7b** shows experimental data of a pulsed LiDAR working under different artificial fog densities, simulating different meteorological visibilities. One can see that stochastic models (**Figure 7a**) reproduce the expected behavior of light (**Figure 7b**).

Using the simulations and knowing the specifications of our sensor, one can adjust electronic gain and optical power to avoid saturation of it, or even come up with solutions related to the optical design of the system.

Next, in **Figure 8** we present the temporal profiles under the same conditions as in **Figure 7a**, but with the presence of an object placed at 0.5 m with a reflectivity of 100%. If one supposes that the sensor is not saturated under any of the presented conditions, it is observed that around the expected position of the object appears the peak that corresponds to its presence along with the backscattering contribution. Using the simulations, it is possible to evaluate what is the level of scattering at which the target is no longer detected (it is not distinguishable from the backscattering signal), so it would not be possible to obtain its image or compute its distance.

The same type of study can be performed to establish the range limit of the system taken into account between visibility and surface properties of the object. Data from an experimental test are shown in **Figure 9**, and it can be seen how both parameters may influence the results.

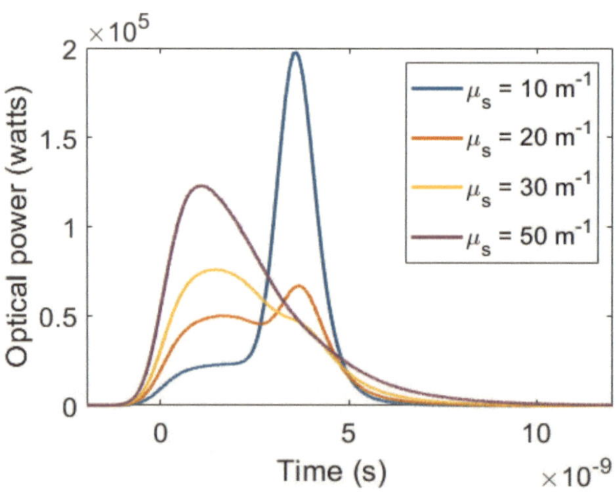

Figure 8.
Detected backscattering in a medium with $g = 0.9$ and different μ_s, for a light pulse with an initial energy of 1 mJ, with the presence of a perfect reflecting target.

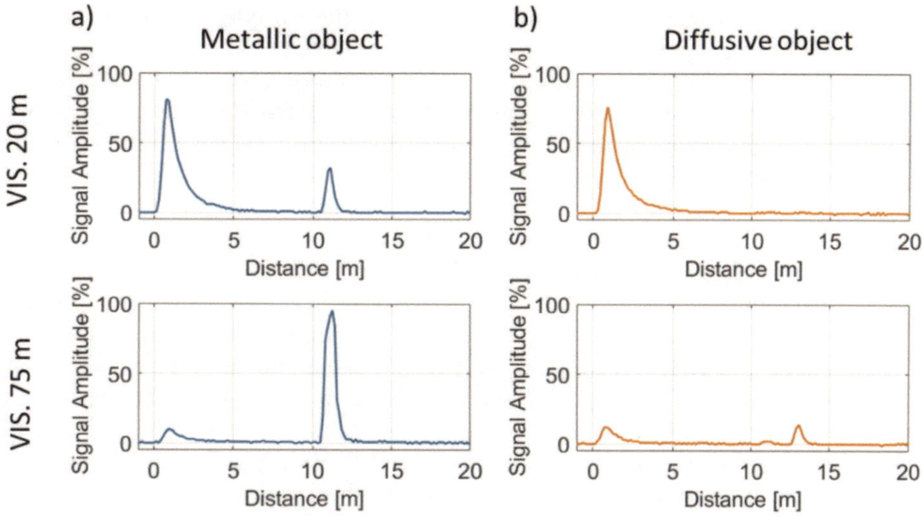

Figure 9.
Signal detected from a returning pulse of light pointing toward (LEFT) a metallic object at 12 m of distance and (RIGHT) a diffusive object at 14 m of distance, for two different visibilities (20 m and 75 m).

5. Conclusions

Over recent years, LIDAR technology has become a *panacea* in the fields of optomechanical engineering and optoelectronics. It especially seems to hold a relevant role in novel applications related to outdoor environments. One of the keys to its success is the amount of information it can provide despite relying on a very simple method. Being based on a simple working principle (counting elapsed time between events in magnitudes carried out by light, e.g., reflected energy from a pulse of light sent to a target), it allows the performance of complete 3D mapping with outstanding characteristics.

However, for the complete settling of the technology, there are still obstacles pending to be solved. One of the most challenging is related to its outdoor performance. As with any other optical sensor, under the presence of adverse weather conditions, such as fog, the system performance is heavily altered and the quality of the detection becomes severely degraded.

Usually, commercial systems are like black boxes, only returning the point cloud or 3D map. When facing bad weather, data are unreliable and its behavior is unknown. However, the LiDAR system can provide a lot more information, which might enable optimized features.

In order to propose reasonable solutions, the knowledge of optical physics behind the phenomena degrading the performance of the system may be of significant interest. The propagation of light through scattering media, such as adverse weather conditions, shows two main problems: the saturation of the sensor and the attenuation of the pulse. Both are related to the dispersive effect and absorption characteristics of the media. Due to their nature, both phenomena can be easily studied with models if the optical properties of the media are known.

Using Mie Theory, and taking into necessary conditions, we can obtain an approximation of the media properties (absorption coefficient, scattering coefficient, phase function, and anisotropy factor). The derivation of the complete theory may be long and tedious; however, its application is straightforward.

Once all the media properties are derived, they can be used to solve the radiation transfer theory. Essentially, it is the basic theory that allows the calculation of light distributions in multiple scattering media with absorption. Its core is the radiation transfer equation, which computes the balance that characterizes the flow of photons in a given volume element. However, the analytical solution to this equation is complicated and often impossible to solve when boundaries, inhomogeneities, or nonstationary effects are involved. The Monte-Carlo (MC) method is used as the conventional tool to arrive at a statistical solution in these cases. By using the MC method, we can solve the RTT of a pulsed LiDAR when working through a turbid medium. Along this chapter, we have shown how the model succeeds in modeling different kinds of scenarios: a media where absorption prevails over scattering, predictions of pulsed light interactions, dynamics of backscattering, light response to different kinds of objects and media, etc. As a result, we are able to predict the behavior of our system in the different scenarios where it breaks down. For example, it is possible to estimate the range of the system, the response toward different objects or the characteristics of the blinding backscattering.

The next step would be related to studying other features that are considered very relevant when designing LiDAR imaging systems, for example:

- Numerical characterization of the backscattering phenomena of active illuminators due to the first interaction with the environment.

- Scattering effects in propagation and its wavelength dependence.

- Optimal configuration sensing to improve contrast-to-noise ratio for imaging (CNR) and radiometric detection.

- Using simulations to find other properties that could be taken into account to discern between signal photons and backscattering photons (for example polarization).

Moreover, one has to not lose focus on the final implementation of the system. Simulations can be used to guide us, to try exotic ideas without the need to build up the system. However, experimental results are always expected as the final product of this whole process.

Acknowledgements

Authors thank Agència de Gestió d'Ajuts Universitaris i de Recerca for Grants 2021FI_B2 0007. Authors also acknowledge support received from DSTL under contract DSTLX1000145661 and the one from Misnitry Sicience and Innovation through Projects PID2020-119484RB-I00 and PDC2021-121038-I00.

Conflict of interest

The authors declare no conflict of interest.

Author details

Maria Ballesta-Garcia[1*], Gerard DeMas-Giménez[1] and Santiago Royo[1,2]

1 Centre de Desenvolupament de Sensors, Instrumentació i Sistemes, Universitat Politècnica de Catalunya (UPC-CD6), Terrassa, Spain

2 Beamagine SL, Castellbisbal, Spain

*Address all correspondence to: maria.ballesta.garcia@upc.edu

References

[1] Royo S, Ballesta-Garcia M. An overview of lidar imaging systems for autonomous vehicles. Applied Sciences. 2019;**9**:4093

[2] Himmelsbach M, Lüettel T, Wuensche HJ, et al. Real-time object classification in 3D point clouds using point feature histograms. In: Proceedings-IEEE/RSJ International Conference on Intelligent Robots and Systems, St. Louis: IROS. 2009. pp. 994-1000

[3] Hecht J. Lidar for Self-Driving Cars. Optics Photonics News. 2018;**29**:26-33

[4] Koopman P, Wagner M. Challenges in autonomous vehicle testing and validation. SAE International Journal of Transportation Safety. 2016;**4**:15-24

[5] Peynot T, Underwood J, Scheding S. Towards reliable perception for unmanned ground vehicles in challenging conditions. In: Proceedings-IEEE/RSJ International Conference on Intelligent Robots and Systems. 2009. pp. 1170-1176

[6] Heinzler R, Schindler P, Seekircher J, Ritter W, Stork W. Weather influence and classification with automotive lidar sensors. In: IEEE Intelligent Vehicles Symposium (IV), Paris. 2019. pp. 1527-1534

[7] Jokela M, Kutila M, Pyykönen P. Testing and validation of automotive point-cloud sensors in adverse weather conditions. Applied Sciences. 2019;**9**:2341

[8] Kutila M, Pyykönen P, Ritter W, et al. Automotive LIDAR sensor development scenarios for harsh weather conditions. In: Proceedings-IEEE 19th International Conference on Intelligent Transportation Systems (ITSC). Rio de Janeiro. 2016. pp. 265-270

[9] Kutila M, Pyykönen P, Holzhuter P, et al. Automotive LiDAR performance verification in fog and rain. In: Proceedings-IEEE Conference on Intelligent Transportation Systems. Maui. 2018. pp. 1695-1701

[10] Ballesta-Garcia M, Peña-Gutiérrez S, Rodríguez-Aramendía A, et al. Analysis of the performance of a polarized LiDAR imager in fog. Optics Express. 2022;**30**: 41524

[11] Woktanowski J, Zygmunt M, Kaszczuk Z, et al. Comparison of 905 nm and 1550 nm semiconductor laser rangefinders' performance deterioration due to adverse environmental conditions. Opto-Electronics Review. 2014;**22**:183-190

[12] Duthon P, Colomb M, Bernardin F. Light transmission in fog: The influence of wavelength on the extinction coefficient. Applied Sciences. 2019;**9**:2843

[13] Ijaz M, Ghassemlooy Z, Minh H, et al. Analysis of fog and smoke attenuation in a free space optical communication link under controlled laboratory conditions. In: IEEE International Workshop on Optical Wireless Communications (IWOW). Pisa. 2012. pp. 1-3

[14] Yoneda K, Suganuma N, Yanase R, et al. Automated driving recognition technologies for adverse weather conditions. IATSS Research. 2019;**43**: 253-262

[15] Heinzler R, Piewak F, Schindler P, et al. CNN-Based Lidar Point Cloud De-Noising in Adverse Weather. In: IEEE Robotics and Automation Letters. Vol. 5. No. 2. 2020; pp. 2514-2521

[16] Lee U, et al. EureCar Turbo: A self-driving Car that can handle adverse

weather conditions. In: 2016 IEEE/RSJ International Conference on Intelligent Robots and Systems (IROS). Daejeon: 2016. pp. 2301-2306

[17] Tuchin V. Tissue optics and photonics: Light-tissue interaction. Journal of Biomedical Photonics Engineering. 2016;**2**:030201

[18] Chandrasekhar S. Radiative Transfer. New York: Dover Publications; 1960

[19] Mishchenko MI. Gustav Mie and the fundamental concept of electromagnetic scattering by particles: a perspective. Journal of Quantitative Spectroscopy and Radiative Transfer. 2009;**110**(14-16): 1210-1222

[20] Mishchenko MI. Gustav Mie and the fundamental concept of electromagnetic scattering by particles: A perspective. Journal of Quantitative Spectroscopy and Radiation Transfer. 2009;**110**:1210-1222

[21] Qiu C, Gao L, Joannopoulos JD, et al. Light scattering from anisotropic particles: Propagation, localization, and nonlinearity. Laser & Photonics Reviews. 2010;**4**:268-282

[22] Boas DA, Pitris C, Ramanujam N, editors. Handbook of biomedical optics. CRC Press; 2016

[23] Saleh BEA, Teich MC. Polarization optics. In: Fundamentals of Photonics. John Wiley & Sons, Inc; 2007. pp. 199-242

[24] Colomb M, Duthon P, Laukkanen S. Characteristics of Adverse Weather Conditions. 2017. p. 73

[25] Hulst HC, van de Hulst HC. Light scattering by small particles. Courier Corporation. 1981

[26] Deirmendjian D. Scattering and polarization properties of water clouds and hazes in the visible and infrared. Applied Optics. 1964;**3**:187

[27] Gebhart M, Leitgeb E, Sheikh Muhammad S, et al. Measurement of light attenuation in dense fog conditions for FSO applications. Atmospheric Optical Model Measurement Simulation. 2005;**5891**:58910K

[28] Metropholis N, Ulam S. The Monte Carlo method. Journal of the American Statistical Association. 1949;**44**:335-341

[29] Whitney B. Monte Carlo Radiative Transfer. In: Saikia DJ, editor. Fluid Flows To Black Holes: A Tribute to S. Chandrasekhar on His Birth Centenary. World Scientific. 2011. pp. 151-176

[30] Jacques SL. Modeling tissue optics using Monte Carlo modeling: A tutorial. Proc-SPIE Opt Interact with Tissue Cells XIX. San Jose. 2008. 68540T

[31] Watté R, Aernouts B, Van Beers R, et al. Modeling the propagation of light in realistic tissue structures with MMC-fpf: A meshed Monte Carlo method with free phase function. Optics Express. 2015;**23**:17467

[32] Jurovata D, Kurnatova J, Ley S, et al. Simulation of photon propagation in tissue using Matlab. Research Papers Faculty of Material Science and Technology. 2013;**21**:31

[33] Prahl SA, Keijzer M. A Monte Carlo model of light propagation in tissue. Proc-SPIE Dosimetry Laser Radiation in Medicine and Biology. Berlin. 1989;**1989**: 1030509

[34] Pattanaik SN, Mudur SP. Computation of global illumination by Monte Carlo simulation of the particle model of light. In: Proc 3th Eurographics Work Render. Bristol. 1992. pp. 71-84

[35] Jacques SL. Monte Carlo modeling of light transport in tissue (steady state and time of flight). In: Optical-thermal response of laser-irradiated tissue. Dordrecht: Springer; 2010. p. 109-144

[36] Carter LL, Cashwell ED. Particle-transport simulation with the Monte Carlo method. United States. 1975. DOI: 10.2172/4167844

[37] Rao KD, Patel HS, Jain B, et al. Time-gated optical imaging through turbid media using stimulated Raman scattering: Studies on image contrast. Journal of Physics. 2005;**64**:229-238

[38] Jacques SL. Light distributions from point, line and plane sources for photochemical reactions and fluorescence in turbid biological tissues. Photochemistry and Photobiology. 1998; **67**:23-32

[39] Kienle A, Patterson MS. Improved solutions of the steady-state and the time-resolved diffusion equations for reflectance from a semi-infinite turbid medium. Journal of the Optical Society of America A. 1997;**14**:246-254

[40] Satat G, Tancik M, Raskar R. Towards photography through realistic fog. IEEE International Conference on Computational Photography. Pittsburgh: ICCP 2018; 2018. pp. 1–10